Atlas of Root Drugs from AYURVEDA

THE AUTHORS

Dr. Sharad Srivastava is Senior Principal Scientist & Head, at Pharmacognosy Division, CSIR-National Botanical Research Institute; Lucknow, INDIA. He has made significant contributions in quality control of crude drugs/products, chemotaxonomy, bio-prospection and natural product development. He had developed quality parameters of more than 70 single crude drugs (medicinal plants) and also identified biomarkers for their quality control. He has contributed 30 monographs of single herbal drugs in Ayurvedic Pharmacopoeia of India. He has 175 publications in peer reviewed journals, 18 patents and developed some technologies/formulations, nine has already been transferred to industry.

Dr. Ankita Misra is working as Young scientist (DST, SYST) at Pharmacognosy Division, CSIR-National Botanical Research Institute, Lucknow (India). She has gained expertise in Pharmacognosy & Phytochemistry of medicinal plants, focussing on chemotaxonomy, analytical chemistry of plant metabolites, and herbal product development. She published 45 research papers, 2 book/chapters and has 2 Indian patents.

Dr. Bhanu Kumar is Senior Research Fellow at Pharmacognosy Division, CSIR-National Botanical Research Institute, Lucknow (India). He is working in the area of pharmacognosy of medicinal plants, chemotyping, quality control of herbal drugs and herbal product development. Published 19 research papers in peer reviewed journals and three book chapters. He received "Popular Science Writing Award" under the scheme of AWSAR by DST, Govt. of India in year 2019-20.

Mr. Mridul Kant Chaudhary is working as UGC-SRF at Pharmacognosy Division, CSIR- National Botanical Research Institute, Lucknow (India). He is working on the botanical studies and quality control of medicinal plants and their chemotaxonomy. He has published 4 research papers in peer reviewed journals.

Atlas of Root Drugs from AYURVEDA

Authors

Sharad Srivastava

Ankita Misra

Bhanu Kumar

Mridul Kant Chaudhary

2022

Daya Publishing House®

A Division of

Astral International Pvt. Ltd.

New Delhi – 110 002

ISBN 978-93-5461-138-4

Published by : **Daya Publishing House®**
A Division of
Astral International Pvt. Ltd.
– ISO 9001:2015 Certified Company –
4736/23, Ansari Road, Darya Ganj
New Delhi-110 002
Ph. 011-43549197, 8130496929
E-mail: info@astralint.com
Website: www.astralint.com

Digitally Printed at : **Replika Press Pvt. Ltd.**

वै.औ.अ.प. – राष्ट्रीय वनस्पति अनुसंधान संस्थान
CSIR - National Botanical Research Institute
राणा प्रताप मार्ग, लखनऊ – 226 001, उ.प्र., भारत
Rana Pratap Marg, Lucknow - 226 001, U.P., India

प्रोफेसर एस. के. बारिक
निदेशक
Professor S. K. Barik
Director

Foreword

The medicinal plants have been an integral part of the history and culture of India, dating back to thousands of years. The deep understanding and wisdom of our ancestors in medicinal plants have given numerous herbal drugs to serve the humanity. The traditional Indian Systems of Medicine (ISM) make use of this knowledge in developing health care products for promoting health and well-being.

Medicinal plants are sources of various secondary metabolites, although the cells/tissues of their synthesis and accumulation vary from species to species. Roots are important source for some well-known Ayurvedic drugs such as Ashwgandha (Withania somnifera (L.) Dunal) and Sarpgandha (Rauvolfia serpentina L.). Among different plant parts which are used for drugs, roots constitute 30-40 % of the total herbal drugs. Being the source of valuable metabolites, roots are always on demand in the herbal drug industry. This results in indiscriminate uprooting of medicinal plants from the wild causing serious depletion of natural populations. Such activities are also responsible for putting valuable medicinal plants into the verge of extinction. Scarcity of genuine raw material and adulteration of authentic raw material with spurious or substandard quality material leads to the poor quality of finished health care products. In the current scenario of increasing demand for herbal drugs and raw material, it is crucial to properly identify authentic samples and establish the quality parameters of raw materials, especially in the case of root drugs.

A consolidated set of data on the botanical and chemical profiles of root drugs is essential for quality control of health care products. Such data are also required for raising awareness among all the stakeholders in academia and industry to avoid unethical practices in herbal drug production sector. The "Atlas of root drugs from Ayurveda" provides chemical and botanical profile of fifty commercially valuable root drugs. It is visually resourced with microscopic illustrations, HPTLC profiles and marker compounds. I am sure that the book would serve as a reference for the scientists and researchers of Botany, Pharmacy and Chemistry, as well as medicinal plant farmers and pharmaceutical industry personnel.

(S.K. Barik)

Phones : (Off.) (0522) 2205848, EPABX Phone : (0522) 2297800 to 2297999 Fax : (0522) 2205839, Post Box No. 436, Lucknow-226001
e-mail : director@nbri.res.in, sarojkbarik@gmail.com, Website : http://www.nbri.res.in

Preface

"Everything is theoretically impossible, until it is done."

– *Robert A. Heinlein*

The traditional systems of medicine are integral part of the glorious heritage of India and *Ayurveda* is one of the most ancient Indian systems of medicine, known worldwide. In this classical text variety of plants and/or plant part are mentioned for its usefulness and medication. In ancient times, the methods for preparation of drug/formulation from the medicinal plants were very precise and decided as per the requirements of individuals. Collection of required medicinal plants was done by *Vaidyas* in a précised manner following all the ethical and scientific methodologies. However, in today's scenario the principles of *Ayurveda* remains the same but procurement of raw materials for development of quality herbal drugs/formulations at industrial scale is facing several challenges. One of these challenges is unavailability of authentic and good quality raw material in herbal drug market, which is due to malpractice of adulteration and substitution to authentic drugs with inferior quality or sometimes altogether different plants or plant parts. Beside this, the other major reason for unavailability of quality raw material is the scarcity of authentic raw material in natural biodiversity due to pressure of constantly increasing global demand of herbal products. Various parts of a plant such as roots, stems, bark, leaves, flowers, seeds *etc.* are used in herbal products and the issue of adulteration and, substitution exists equally, however root drugs are vulnerable due to the morphological similarity with each other.

Out of approximately 900 species commercially used in the Indian systems of medicine, nearly 150 species are root drugs. The root drugs are underground part of plant species and large scale harvesting of sample will pose threat to the biodiversity of the species, except for cultivated one. According to National Medicinal Plant Board (NMPB), Government of India, about 47 per cent of the total medicinal plant species are procured by destructive harvesting methods, especially for root/rhizome/tuber/whole plant drugs.

The present atlas is an attempt to resolve such issue in raw drugs where root part is used for medicinal benefit. This includes 50 raw drugs or "root drugs", judiciously selected due to their traditional value and, high commercial significance and is structured in the form of an atlas. Comprehensive information about each root drug includes its salient morphological and anatomical characteristics for the identification in field or market and under microscope at lab. Along with this a qualitative chemical profile along with reference marker was also included to confirm its authenticity and quality. Anatomical sections are prepared and with the help of microphotographs, the necessary gross structural spectrum of the roots is provided to facilitate in understanding the non science background students also. The High Performance Thin Layer Chromatography (HPTLC) profiles with marker compound(s) was included to ensure the availability of specific phytomolecules in the individual root drug which can play a crucial role in the quality control of raw material. The HPTLC protocol used for analysis in each drug has been mentioned in detail. We believe that this " Atlas of root drugs from Ayurveda" is not limited to undergraduate and postgraduate pharmacy and botany students, but also useful to the teachers, researchers, farmers, traders, plant collectors and scientists working in the area of medicinal plants and drug development.

The authors express their deep sense of gratitude to Prof. Saroj Kanta Barik, Director, CSIR-National Botanical Research Institute, Lucknow for his constant support and encouragement throughout the course activity. Thanks are due to all the lab members, technical and non technical staff for their cooperation and support in the compilation of this atlas.

Sharad Srivastava

Ankita Misra

Bhanu Kumar

Mridul Kant Chaudhary

Contents

Abbreviations

Plant Tissues

cc Calcium oxalate crystal
cjt Conjunctive tissue
ck Cork
cr Crystals
ct Cortex
ctvb Cortical vascular bundle
end Endodermis
epi Epidermis
gt Ground tissue
ls Longitudinal section
mr Medullary rays
mx Metaxylem
od Oil droplets
orc Oleoresin canals
ph Phloem
phf Phloem fibre
pi Pith
px Primary xylem
pxy Proto xylem
rt Root primordium
sc Sclereids
sg Starch grains
stc Stone cells
stvb Stealer vascular bundle
sx Secondary xylem
trd Tracheid
ts Transverse section
vb Vascular bundle
ve Vessels
xy Xylem
xyf Xylem fibre
xyv Xylem vessels

Technical Terms

µl Microlitre
°C Degree celsius
± Statistical margin
≥ Less or equal
cm Centimetres
ft Feet
g Gram
h Hours
HPTLC High performance thin layer chromatography
m Meters
mg Milligram
min Minute
ml Millilitre
mm Millimetre
N Normal
nm Nano meter
R Reference
R_f Retention factor
S Sample
UV Ultra violet
v/v Volume per volume

Introduction

"Establishment of a healthy & prosperous society can be attained by more holistic approach - Medicinal Plants"

Herbal medicines continue to be a popular healthcare choice for the general public not only for health maintenance and wellbeing, minor ailments and serious chronic condition, but also for 'enhancement' of functions or processes (Barnes *et al.*, 2007). Natural products contribute a vital role in the discovery of new pharmacologically active compounds, which are clinically useful drugs to cure diseases. Herbs provide bio-active products as well as the starting material for the new synthetic or semi- synthetic drug entity. Thus, the morden world is diverted towards safe alternatives for treatment of diseases and thus, herbal medicines are one of the best choices till date. However, the global concern has been focused to herbal medicines that have scientific authenticity for their efficacy, toxicity and safety.

One of the impediments in the acceptance of the herbal products worldwide is the lack of standard quality control profiles. Quality control of traditional medicines involve the quality of both raw materials, mostly plants but also animals, metals and minerals, as well as of finished products. It is very important to be aware that the primary identity of the material can only be decided by a reliable traditional source. Subsequent to this, identification, morphological, microscopically, chemical and biochemical methods are available to undertake standardization of traditionally identified material and are appropriately applied. ***"Quality can be defined as the status of a drug which is determined by identity, purity, content, and other botanical, physical, chemical, biological or toxicological properties, or by the manufacturing processes"***. The word quality has many meanings (1) a degree of excellence, (2) conformance with requirements, (3) the totality of characteristics of an entity that bear on its ability to satisfy stated or implied needs, (4) fitness for use, (5) fitness for purpose, (6) freedom from defects, imperfections or contamination and (7) delighting customers. It describes all measures taken during manufacturing process which leads to reproducible quality of particular product, as per guidelines laid by official regulatory authorities of various countries and the World Health Organization. The medicinal benefit of plants is distributed in its various parts and depending upon which raw drugs are classified into several categories. Most of the time whole plant is used or/otherwise plant parts such as root, stem, leaf, flower, seed, fruit and modifications of stem and root, bark of a stem or root, wood, and their exudates or gums *etc.* constitute single drugs in the Indian Systems of Medicine (ISM).

Raw Drugs

Human beings are using medicinal plants since ages based on their wisdom and experience. People use different plants and plant parts (root, stem, leaves, flowers, fruits, bark, resins and gums *etc.*) in various dosage forms for maintaining health and treating diseases. Medicinal plants are rich sources of therapeutically important secondary metabolites and different plant parts are sources of various metabolites. Therefore, before plant collection it is important to learn about the bioactive plant part that constitute major part of the raw drug.

Types of Raw Drugs

Raw or crude drugs are available in the commercial markets in one of the following forms:

- ☆ Entire, nearly entire, or broken eg. Leaf, root, fruit *etc.*
- ☆ Matted together condition *e.g.* gum, exudates
- ☆ Transverse slices *e.g.* bark
- ☆ Longitudinal slices *e.g.* stems
- ☆ Cut forms *e.g.* stem,
- ☆ Shredded forms *e.g.* shreaded leaves
- ☆ Peeled conditions *e.g.* fruits/fruit rind
- ☆ Powdered form *e.g.* whole plants/parts in particular

Important Root Drugs

Root drugs are an important class, out of 900 medicinal species being used as indigenous medicine, approximately 150 species are root drugs. There are some potential formulations in *Ayurveda* which constitutes root exclusively as drugs such as *Dashmoolarista.* It contains root of ten medicinal plants *i.e. Desmodium gangeticum, Uraria picta, Tribulus terrestris, Aegle marmels, Oroxylum indicum, Solanum indicum, Solanum surattense, Gmelina arborea, Clerodendrum phlomidis and Steresspermum chelonoides.* One major problem with root drugs is that harvesting at large scale posseses threat to the species biodiversity. In case of high market demand, indiscrimainate collection/harvesting from natural resourses is a common practice along with adulteration or susbstitution with spurious quality material due to profit earning. Some commercially important root drugs are:

- ☆ *Withania somnifera* (Aswagandha)
- ☆ *Glycyrrhiza glabra* (Mulethi)
- ☆ *Asparagus racemosus* (Satavari)
- ☆ *Chlorophytum borivilianum* (safed musali)
- ☆ *Curculigo orchioides* (Kali musali)
- ☆ *Rauwolfia serpenina* (Sarpgandha)
- ☆ *Sida cordifolia* (Bala)
- ☆ *Plumbago zeylanica* (Chithrak mool)
- ☆ *Hemidesmus indicus* (Anant mool)
- ☆ *Berberis aristata* (Daruharidra)

Challenges in Quality Control of Herbal Drugs

There are several challenges in quality control of herbal drugs such as, false identification of the plant source in natural conditions, adulteration of spurious material in the authentic drug, microbial contamination, heavy metal load, pesticides *etc.* All these factors led to change in phytochemical constituents of raw material and perpetually affects the efficacy and safety of the herbal products. Besides, variations in environmental conditions, soil micro-flora and other biotic-abiotic factors affect the raw materials due to change in the concentration of bioactive metabolites. The World Health Organization (WHO) has laid guidelines for the quality control of herbal drugs and emphasized on the need of using modern tools and techniques. The effective implementation of Good Agricultural and Collection Practices (GACP) issued by WHO can be very helpful in addressing many issues related to quality of herbal drugs.

Adulteration and Substitution

Adulteration is the most common malpractice which leads to the inferior quality of herbal product. Adulteration is a practice of mixing to the original crude drug partially or fully with other substances. It may also be defined as mixing or substituting the original drug material with other spurious, inferior, defective, spoiled, useless other parts of same or different plant or harmful substances or drug which do not confirm with the official standards.

Adulteration, however, is the biggest challenge in herbal drug industries and herbal products as many crude drugs are prone for adulteration due to lack of standard parameters for their identity. Beside this the other factors provoking this practice is lack of species biodiversity in collection, earning of more profit, lack of scientific knowledge *etc.*

Tools for Identification of Raw Drugs

The identification of raw drugs can be carried out by evaluating the organoleptic characters, anatomical studies and chemical characters as mentioned below:

1. Organopeltic Characters

Analysis of organoleptic characters is an important tool in the quality control of herbal drugs. It is particularly significant in case of root drugs as these drugs are morphologicaly indistinguishable in most of the cases. Organoleptic characters are evaluated on the basis of shape, size, colour, odour, taste and fracture of raw drugs *etc.* The organoleptic evaluation done by means of sensory organs is the preliminary method to know about the identity and quality of a particular drug *e.g.*, *Berberis* root can be identified by its bright yellow colour while *Hemidesmus* can be identified with its characteristic aroma.

2. Anatomy

Plant anatomy is an integral component of plant taxonomy and an essential tool in Pharmacognosy. The anatomical perspective of plant organ offers a unique botanical diagnosis of fragmentary plant materials and plays a significant role in vascular plant systematics. In many instances, anatomical studies of plant materials have provided valuable clues in resolving plant identification issues *e.g.*, identification on the basis of presence or absence of specific cells or structures such as, arrangement of vascular bundles.

3. Chromatography

Thin Layer Chromatography (TLC)

Thin layer chromatography (TLC) is a method for separating the compounds from a mixture. TLC can be used to determine the number of components in a mixture, the identity of compounds, and the purity of a compound. TLC consists of three steps – sample application, development and visualization. The technique has advantage than others' due to its precision, accuracy yet low cost of analysis. TLC can be qualitative and quantitative. The fingerprinting profile developed after analysis can be used as qualitative, when only the presence and absence of particular known and unknown metabolites are identified. In quantitative estimation the presence of a particular metabolite can be quantified by using reference markers. TLC has achieved phenomenol success not only for its use in analytical problems but also in preparative chromatography. The retension factor (Rf) is used to indicate the movement of phytomolecule along the plate with the aid of solvent system. Rf is equal to the distance traveled by substance to the distance traveled by the solvent. Its value is always between zero and one.

High Performance Thin Layer Chromatography (HPTLC)

An extension of TLC, high-performance thin layer chromatography (HPTLC) is robust, simplest, rapid, and efficient tool in quantitative analysis of compounds. HPTLC is an advanced versatile chromatographic technique for quantitative analyses with high sample throughput and is complementary to HPLC/GLC. It provides a chromatographic fingerprint and is therefore suitable for monitoring the identity and purity of drugs. In HPTLC, the plates are coated with silica gel which is of uniform size (about 5 mm). These silica gel precoated plates give more efficient and reproducible separation than conventional grades of silica gel. HPTLC is an analytical technique based on TLC, but with enhancements intended to increase the resolution of the compounds to be separated and to allow quantitative analysis of the compounds. Some of the enhancements such as the use of higher quality TLC plates with finer particle sizes in the stationary phase which allows better resolution. Beside this, the separation of metabolite is under the controlled condition of temperature, pressure and relative humidity.

Unlike other plant part roots are one among the plant materials that need simple but effective standards of microscopic features that can be readily employed during exigence. The primary scope of the present study is to analyse the important structural features and chemical characterization of some selected root drugs.

Materials and Methods

Roots of 50 medicinal plants belonging to different families were selected for the microscopic and HPTLC characterization. The plant from which roots are harvested, were authenticated with the help of various floras and herbarium specimen's available in our National repository, CSIR-NBRI. The medicinal properties of the roots were retrieved from information available in public domain, both online and off line literature (Chopra *et al.*, 1956). The enlisted (Table 1) species are commercially relevant in herbal drug industry but consolidated quality control and identification keys are missing or scattered. Therefore, the targeted root drugs were selected in this atlas.

Table 1: of plants for anatomical and chemical characterization

Sl.No.	Plant Name	Family
1.	*Abutilon indicum* Linn.	Malvaceae
2.	*Acorus calamus* Linn.	Acoraceae
3.	*Achyranthes aspera* Linn.	Amaranthaceae
4.	*Aristolochia indica* Linn.	Aristolochiaceae
5.	*Berberis aristata* DC.	Berberidaceae
6.	*Berberis asiatica* Roxb.	Berberidaceae
7.	*Berberis chitria* Lindl.	Berberidaceae
8.	*Berberis lycium* Royle	Berberidaceae
9.	*Berberis tinctoria* Lesch.	Berberidaceae
10.	*Berberis umbellata* Wall. ex G. Don	Berberidaceae
11.	*Bergenia ciliata* (haw.) Sternb	Saxifragaceae
12.	*Bergenia ligulata* Wall	Saxifragaceae
13.	*Bergenia stracheyi* (Hook.f. & Thoms.)	Saxifragaceae
14.	*Boerhaavia diffusa* Linn.	Nyctaginaceae

Sl.No.	Plant Name	Family
15.	*Calotropis gigantea* (L.) R. Br.	Asclepiadaceae
16.	*Calotropis procera (Ait.) R.Br.*	Asclepiadaceae
17.	*Coleus forskholii* Briq.	Lamiaceae
18.	*Coptis teeta* Wall.	Ranunculceae
19.	*Costus speciosus* (Koen. ex Retz.)	Costaceae
20.	*Curcuma amada Roxb.*	Zingiberaceae
21.	*Curcuma angustifolia* Roxb.	Zingiberaceae
22.	*Curcuma aromatica* Salisb.	Zingiberaceae
23.	*Curcuma longa* linn.	Zingiberaceae
24.	*Curcuma zedoaria* Rosc.	Zingiberaceae
25.	*Cyperus rotundus* linn.	Cyperaceae
26.	*Desmodium gangeticum* (L.) DC.	Fabaceae
27.	*Dioscorea deltoidea Wall. Ex Griseb.*	Dioscoreaceae
28.	*Elephantopus scaber* Linn.	Asteraceae
29.	*Gloriosa superba* Linn.	Colchicaceae
30.	*Hemidesmus indicus* (L.) R.Br.	Apocynaceae
31.	*Ichnocarpus frutescens* Linn.	Apocynaceae
32.	*Mahonia borealis* Takeda	Berberidaceae
33.	*Mahonia leschenaultii* (Wallich ex Wight & Arn.) Takeda	Berberidaceae
34.	*Mahonia nepaulensis* DC.	Berberidaceae
35.	*Plumbago zeylanica* Linn.	Plumbaginaceae
36.	*Podophyllum hexandrum* Royle	Berberidaceae
37.	*Potentilla fulgens* L.	Rosaceae
38.	*Pueraria tuberosa* (Roxb. Ex Willd.) DC.	Fabaceae
39.	*Rauvolfia serpentina* L. Benth. Ex Kurz.	Apocynaceae
40.	*Ricinus communis* Linn.	Euphorbiaceae
41.	*Rubia cordifolia* Linn.	Rubiaceae
42.	*Sida cordifolia* Linn.	Malvaceae
43.	*Sida rhombifolia* Linn.	Malvaceae
44.	*Solanum incanum* Linn.	Solanaceae
45.	*Solanum indicum* Linn.	Solanaceae
46.	*Solanum virginianum* L. Syn. *Solanum xanthocarpum* Schrad. and Wendl.	Solanaceae
47.	*Tephrosia purpurea* (L.) Pers.	Fabaceae
48.	*Thalictrum foliolosum* DC.	Ranunculaceae
49.	*Tinospora cordifolia* (Thunb.) Miers	Menispermaceae
50.	*Utleria salicifolia* Bedd. Ex. Hook. F.	Periplocaceae

Sample Preparation for Anatomical Study

Fresh and healthy plants were collected and the roots were carefully removed after moisturing and digging the soil. Photograph of the external profile of the root was taken from the fresh specimens. Small segments of roots with different thickness were cut with sharp razor and fixed immediately in 70 per cent alcohol. After 24 hours of fixing, the specimens were dehydrated employing graded series of Tertiary Butyl Alcohol (TBA) as per the schedule given by Sass (1940) and infiltrated with paraffin wax (melting point 56 – 58 °C). Materials embedded in wax blocks were sectioned to the thickness of 10 – 12 µm with rotary microtome. After dewaxing, the sections were stained with *toluidine blue O* following the procedure of O' Brien *et al.* (1960). Canada Balsam was used as the mounting medium. Toluidine blue is a polychromatic stain and imparts different colours to different tissues depending upon the chemical nature of the cells. Maceration of xylem element was carried out using Jeffrey's maceration fluid (Johansen, 1940). Macerated tissue was stained with 0.5 per cent Safranin, mounted on a drop of glycerine and studied. The sections were photographed with NIKON lab Photo-2 photomicrographic unit at different magnifications. Magnifications of the micrographs are indicated by the scale-bars. Measurements of cells were calibrated using eye-piece micrometer. The descriptive terms are given as per the recommendations of Trockenbroadt (1990) and IAWA Multilingual Glossary (1964).

HPTLC Instrumentation

1. **Stationary phase:** Pre-coated silica gel 60 F_{254} plate (Merck) of uniform thickness 0.2 mm.
2. **Preparation of methanolic extract:** Methanolic extract of sample was prepared through cold maceration by using fixed quantity (5 gm) of powdered material in methanol (HPLC grade). Modification if any, in the extraction protocol is mentioned wherever required.
3. **Sample preparation:** A stock solution of methanolic extract of concentration 10 mg/ml was prepared from the above methanolic extract of sample.
4. **Reference preparation:** A stock solution of concentration 1 mg/ml was prepared in HPLC grade methanol for reference standard.

The stock solutions were stored at 4°C until analysis. Each day, aliquots of the stock solution (standards) were diluted in 10 ml volumetric flasks with methanol to prepare a working solution of 0.1 mg/ml. Samples were prepared by dissolving the known amount of extract in methanol to obtain a final concentration of 10 mg/ml on the day of analysis. Working dilutions of standard and samples were duly filtered through a 0.45 mm Millipore membrane filter (Pall, USA) for HPTLC analysis. All the chemicals and reagents used were of analytical grade.

HPTLC Chromatographic Conditions

Chemical profiling was carried out on 10 cm × 10 cm TLC aluminum precoated plates (normal phase) with 0.2 mm layer thickness of silica gel 60 F_{254} (Merck, Germany). For the quantification, tracks of standard (2 µl) and sample (10 µl) were applied as 8 mm band width using Camag 100 µl sample syringes (Hamilton, Switzerland) with an Automatic TLC sampler (Camag ATS 4, Camag, Switzerland) under a flow of N_2 gas. The linear ascending development was carried out in a Camag glass twin trough chamber. The saturation time of chamber was conditioned and optimized to 5 min at room temperature (25°C ± 2) and relative humidity for better resolution with mobile phase vapors. The plate was allowed to develop up to the height of approximately 85 mm from the point of application (total length run by mobile phase) and the total run time was standardized at room temperature (25°C ± 2) and relative humidity of 55 per cent ± 2. After development, the plates were air dried for 30 minutes and visualized under UV (254 or 366 nm) and/or daylight after derivatization, depending upon the metabolite. The UV spectra and retention factor of standard/ reference marker was compared with sample (Wagner, 1984).

REFERENCES

Barnes J, Anderson LA, Phillipson JD. 2007. Herbal Medicines, Pharmaceutical Press, Edn. 3rd, pp. 8-10.

Cutler DF. 1976. Variation in root-wood anatomy in wood structure in *Biological and Technological Research (eds.* P. Baas A.J., bolton and D.M. Caiiing), Leiden Botanical Series No.3, 143-156 Leiden university Press, Leiden.

Cutler DF, Richardson IBK. 1951. Tree Roots and Buildings, Construction Press, London.

Chopra RN, Nayar SL, Chopra C. 1956. Glossary of Indian Mediclinal Plants. National Institute of Science Communication. CSIR Publication New Delhi lndia. pp.330.

Sass JE. 1940. Elements of Botanical Microtechnique Me Graw Hill Book Co. New York. pp.222.

O'Brien TP, Feder N, Me Cull ME. 1964. Polychromatic staining of Plant cell walls by toluidine blue - O. Protoplasma. 59: 364-373.

Johansen DA. 1940. Plant Microtechnique. Mc Graw Hill Book Co., New York. pp.523.

Trockenbroadt M. 1990. Survey and Discussion of the Terminology used in bark anatomy. IAWA Bull. 11: 141-166.

IAWA committee on Nomenclature. 1964. Multilingual Glossary of Terms used in Wood Anatomy.

Wagner H, Bladt S, Zgainski EM. 1984. Plant drug analysis: a thin layer chromatography atlas. Berlin, New York. Springer-Verlag

1

Abutilon indicum Linn.

Family

Malvaceae

Vernacular Name [1]

Sanskrit: Atibala, Kankatika; Hindi: Kanghi, Jhampi; Bengali: Badela; English: Country mallow, Indian mallow; Marathi: Mudra, Petari; Tamil: Tutti, Nallatutti; Urdu: Kanghi.

Distribution

It is found in the Himalayas ranging from Jammu and Kashmir to Bhutan up to an altitude of 1500 m and spreading through the northern and central India.

Traditional Uses [2]

Their roots are considered as demulcent, diuretic, in chest infection and urethritis; it is also used in fever, haematuria, bronchitis, polyurea and leprosy.

Major Chemical Constituents [3]

β-sitosterol, ephedrine, gallic acid and caryophyllene

Pharmacological Activity [4–6]

Antioxidant, antiulcer and wound healing activity

Plant Description

Abutilon indicum Linn. is a herbaceous plant grows up to 2 meter in height dispersed as a weed on wasteland.

Plant in Natural Habitat

Dried Piece of Root

Leaves are ovate, acuminate, margin dentate, linear, petiole is present and hairy. Stem is yellow coloured, cylindrical, soft, smooth and fibrous. Flowers are yellow in colour, peduncle jointed, axillary, solitary and capsular fruits. The root of the plant is smooth, cylindrical, and fibrous with strong fragrance and salty taste.

Anatomy

Transverse section of root is almost circular in outline. The outer layer is made up of polygonal tubular 4 to 6 layered cork cells, followed by cork cambium. The cortical region is narrow and made up of 10 to 15 or more layered irregular parenchymatous cells; in whole cortical region sclereids are oftenly seen. Sclereids are observed in a group of 5 to 20 or more in numbers. Just below the cortex 4 to 10 layered thick phloem having sieve tubes, companion cells and phloem fibers are present. Medullary rays are 2 – 3 layered radially elongated, parenchymatous cells. The xylem consists of vessels, fibers, tracheids and xylem parenchyma. Vessels are mostly solitary or in groups of 2 or 5.

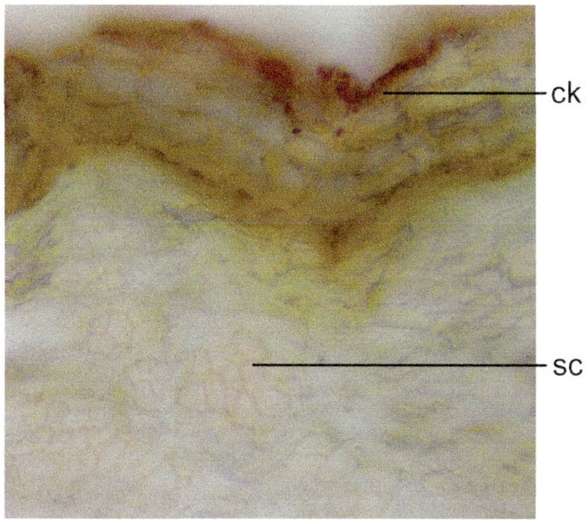

T.S. (40 x)

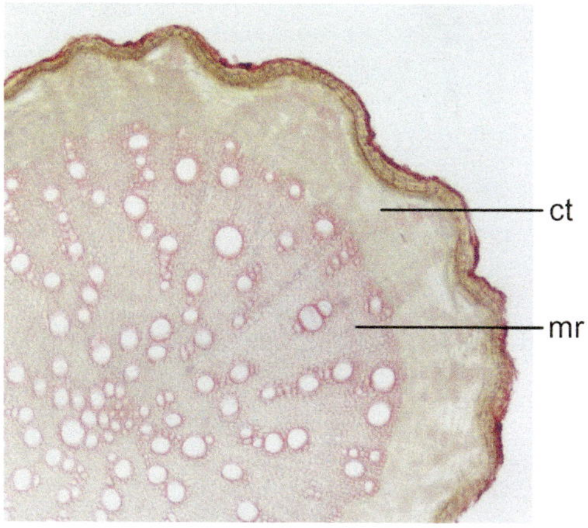

T.S. (4 x)

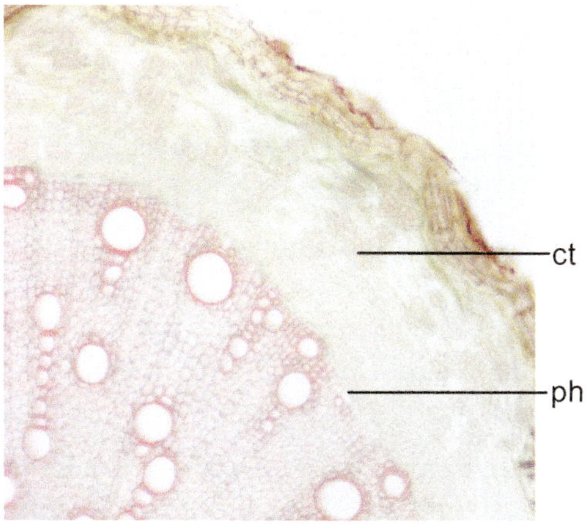

T.S. (10 x)

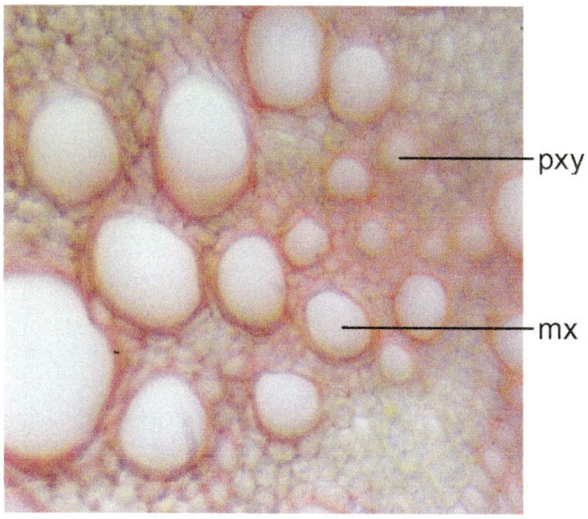

T.S. (20 x)

Transverse Section (T.S.) of *Abutilon indicum* root.

Abbreviation: ck- cork; ct- cortex; mr- medullary rays; mx- metaxylem; ph- phloem; pxy- protoxylem; sc- sclereids.

HPTLC Studies

Working Solution

Accurately weighed (5g) crude powdered sample was subjected to cold extraction with methanol (50 ml), continuously shaken for 6h and then allowed to stand at room temperature (25 ± 2°C) for 18h. Extraction was repeated thrice, and the pooled filtrate was concentrate in rotatory evaporator (Buchi, Switzerland). Working solution for HPTLC study was prepared at a concentration of 10 mg/ml.

Standard Solution

β-sitosterol (> 98 per cent) was freshly prepared at a concentration of 1 mg/ml.

Sample Application

2 μl of standard solution and 10 μl of sample was applied as band by HPTLC applicator on HPTLC preset silica gel GF254 plate. The plate was allowed to developed in CAMAG twin trough glass chamber, up to 8 cm, pre-saturated (15 min) with solvent system.

Solvent System

Toluene: ethyl acetate: methanol: acetone (7:2:0.2:0.2 v/v).

Visualization

The chromatogram was visualized after derivatization with anisaldehyde sulphuric acid reagent and densitometric scanning of β-sitosterol was done at 530 nm.

Identification of β-*sitosterol*

The standard β-sitosterol was identified at R_f 0.55.

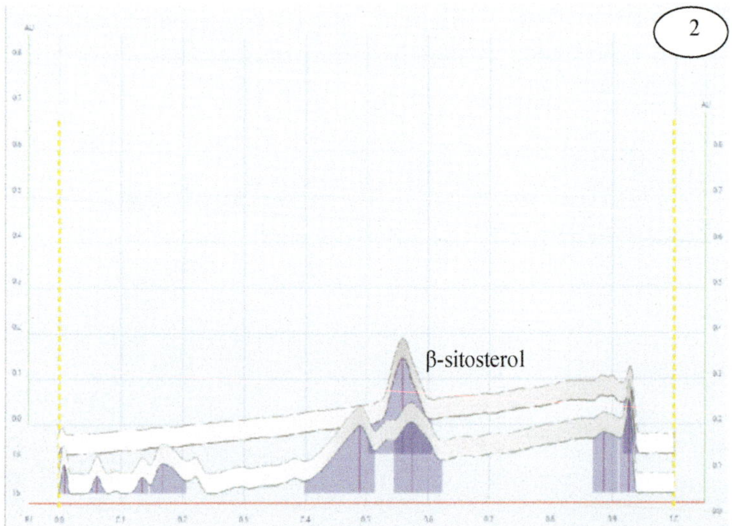

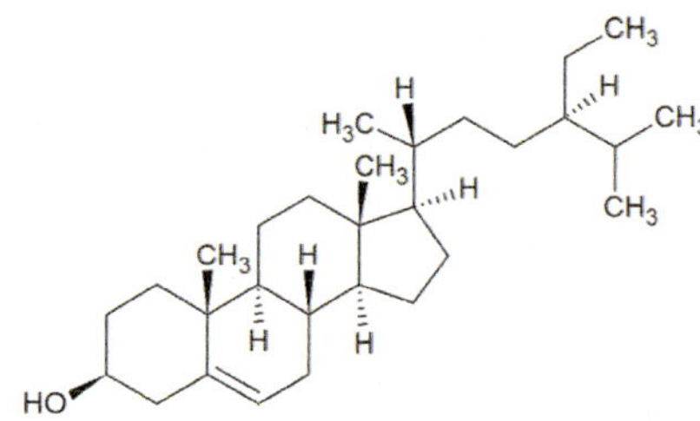

β-sitosterol

Plate 1: HPTLC chromatogram of *Abutilon indicum* sample (S) and standard β-sitosterol (R) visualized after derivatization with anisaldehyde sulphuric acid reagent.

Plate 2: Overlay Spectra of Sample and Standard.

1. Khare CP (2007). Indian medicinal plants, An Illustrated Dictionary. Springer-Verlag Heidelberg, New York 3-4.
2. Chatterjee A, Prakash C (1991). The treatise on Indian Medicinal Plants, Publication and information directorate, New Delhi 174-175.
3. Pengelly A (2004). Triterpinoids and saponins, in the constituent of medicinal plants, CABI publishing, USA 74.
4. Dashputre NL, Naikwade NS (2011). Evaluation of Anti-Ulcer Activity of Methanolic Extract of *Abutilon indicum* Linn Leaves in Experimental Rats. Int J Pharmaceut Sci Drug Res 3(2):97-100.
5. Yasmin S, Kashmiri AM, Asghar MN, Ahmad M, Mohy-ud-Din A (2010). Antioxidant potential and radical scavenging effects of various extracts from *Abutilon indicum* and *Abutilon muticum*. 48(3):282-289.
6. Roshan S, Ali S, Khan A, Tazneem B, Purohit MG (2008). Wound healing activity of *Abutilon indicum*. Pharmacognosy magazine, 4(15):85-88.

2

Acorus calamus Linn.

Family

Acoraceae

Vernacular names

Sanskrit: Bhadra, Bhutanashini, Vacha, Jatila; Hindi: Bach, Ghorbach, Safed bach, Bajai; Bengali: Bach; Marathi: Vekhand; English: Sweet flag, Calamus, Myrtle grass; Gujarati: Vekhand; Telengu: Vadaja, Vasa.

Distribution [1-2]

The plant is found in northern temperate and subtropical regions of Asia, North America and Europe. In India, this plant is growing in wild up to an altitude of 2200 m in the Himalayas. It is plentiful in the marshy area of Kashmir, Himachal Pradesh, Manipur, and Naga hills, and is cultivated in Karnataka and other parts of country.

Traditional Uses [3]

In Ayurvedic medicinal practice, rhizome has been used to cure several diseases like fever, asthma and bronchitis, and as a sedative. Native tribes used it to treat cough, make a decoction as a carminative. Vacha powder with ghee is given ritually in India for improvements in intelligence and speech development. In China it is used in a similar way, to improve speech and aid recovery from stroke.

Vegetative Plant

Fresh rhizome

Major Chemical Constituents [4-5]

β-asarone, camphene, eugenol, calamene, calamenenol and calameone.

Major Pharmacological Activity [6-8]

Antimicrobial, insecticidal, anticancer and immunosuppressive activity.

Morphological Description

It is a perennial herb with creeping and extensively branched, aromatic rhizome grows up to 30 to 50 cm in height. Leaves are simple and contain a single prominent mid-vein, smooth, sub-sessile, elliptic or ovate. Flowering condition is rare, and cylindrical in shape and greenish brown. The fresh rhizome is curved or sometime straight, cylindrical, 2-2.5 cm thick. There are many rootlets arise from entire surface of rhizome.

Anatomy

Transverse section of rhizome is circular in outline, shows single layer of epidermis with wider cortex occupying major area of section, well distinguished endodermal layer and wide stellar system in ground tissue. Detailed structure accounted that the rhizome shows single layer of epidermis, followed by multi-layered cortex. Wide central ground tissue is parenchymatous and stealer vascular bundle is scattered throughout ground tissue. Steles are collateral, non-lignified fibro-vascular bundles; they have been more numerous adjacent peripheral region of endodermis. Stellar vascular bundles are well developed and associated with narrow phloem. Simple oval to pear shaped starch grains are present throughout the ground tissue.

HPTLC Studies

Sample Preparation

Accurately weight (5 g) crude powdered rhizome was extracted with methanol (50 ml) by continuously shaken for 6 h and then allowed to stand for 18 h at

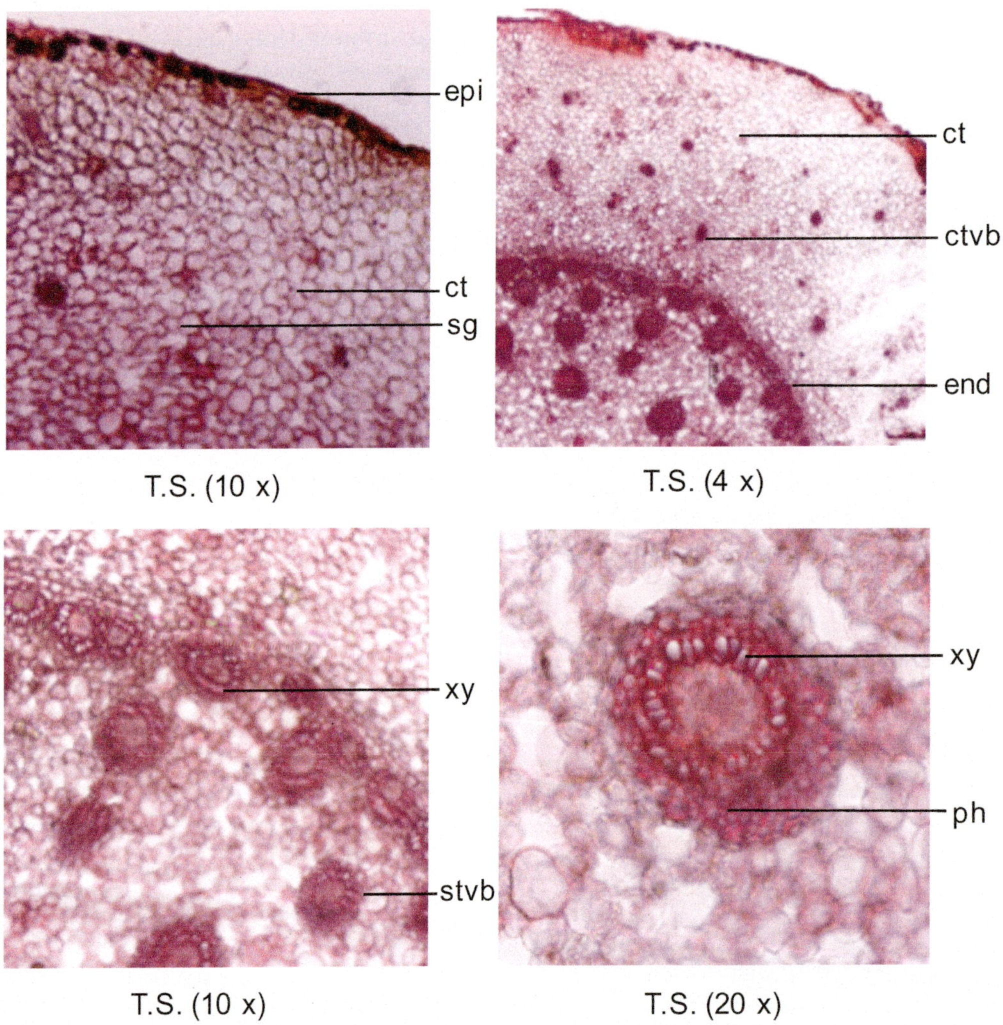

Transverse Section (T.S.) of *Acorus calamus* Rhizome.

Abbreviation: ct- cortex; ctvb- cortical vascular bundle; end- endodermis; epi- epidermis; ph- phloem; sg- starch grains; stvb- stealer vascular bundle; xy- xylem.

room temperature (25°C ± 2). Extraction was repeated thrice, filtered and the pooled filtrate was concentrate in rotatory evaporator (Buchi, Switzerland), dried residue was stored at 4°C. Working solution of 10 mg/ml was freshly prepared on the day of analysis.

Standard Solution

Working solution of β-asarone (≥ 98 per cent) was freshly prepared of strength 1 mg/ml.

Sample Application

2 µl of standard solution and 10 µl of sample was applied as 6 mm band with CAMAG HPTLC applicator on activated silica gel 60 GF_{254} HPTLC plate. The chromatogram was developed on pre-saturated CAMAG HPTLC twin trough chamber for 20 min and the solvent system was allowed to run up to 8 cm.

Solvent System

Toluene: ethyl acetate: diethyl amine (7:2:1 v/v)

Visualization

The chromatogram was visualized at UV_{254} and densitometric scanning of β-asarone was done at 300 nm.

Identification of β-asarone

The standard β-asarone was identified at R_f 0.85.

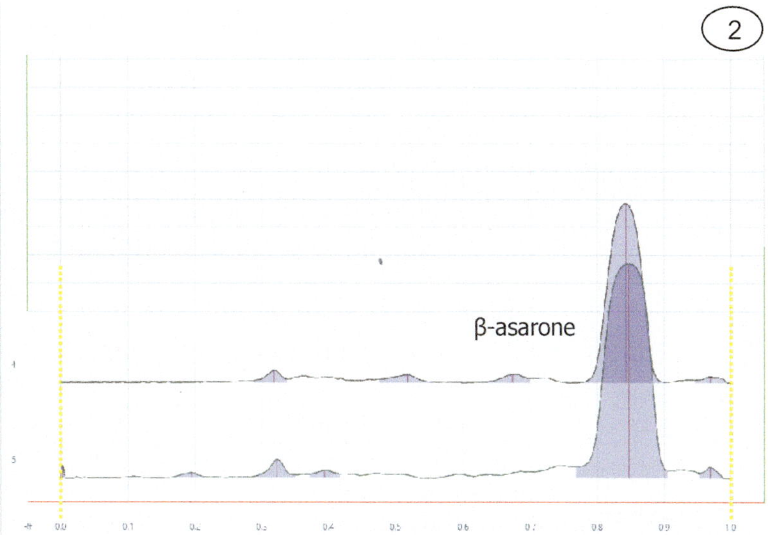

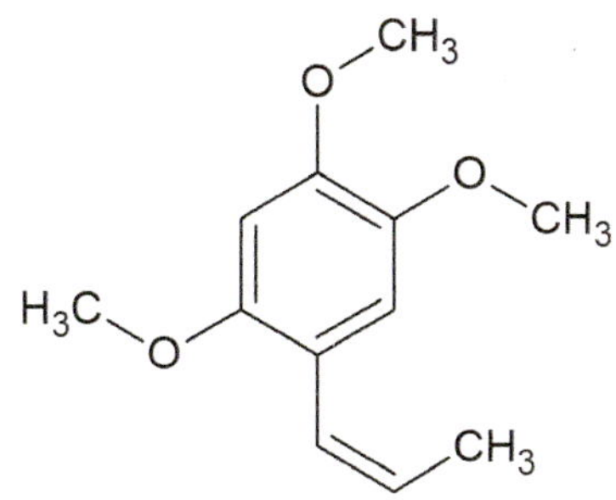

β-asarone

Plate 1: HPTLC Chromatogram of *Acorus calamus* Sample (S) and Standard β-asarone (R).
Plate 2: Overlay Spectra of Sample and Standard.

1. Cavazza G (1976). Les chimiotypes parmi les plantes aromatiques cas du calamus polyploide. Ann Fals Exp Chins 69: 833–844.
2. Anonymous. The wealth of India, Vol. 1. New Delhi: Council of Scientific and Industrial Research; 2004. p. 19 65.
3. Howes, M.R., Houghton, P.J (2003). Plants used in Chinese and Indian traditional medicine for improvement of memory and cognitive function. Pharmacology Biochemistry and Behavior 75, 513–27.
4. Chaudhury SS, Gautam SK, Handa KL (1957). Composition of calamus oil from calamus roots growing in Jammu and Kashmir. Indian J Pharm 19: 183–186.
5. Raja AE, Vijayalakshmi M, Devalarao G 2009. *Acorus calamus* linn: Chemistry and Biology. Res J Pharm Tech 2: 256 61.
6. Phongpaichit S, Pujenjob N, Rukachaisirikul V, Ongsakul M (2005). Antimicrobial activities of the crude methanol extract of *Acorus calamus* Linn. Songklanakarin J. Sci. Technol 27(2): 517-523.
7. Bains JS, Dhuna V, Singh J, Kamboj SS, Nijjar KK, Agrewala JN (2005). Novel lectins from rhizomes of two *Acorus* species with mitogenic activity and inhibitory potential towards murine cancer cell lines. Int Immunopharmacol 5: 1470–1478.
8. Mehrotra S, Mishra KP, Maurya R, Srimal RC, Yadav VS, Pandey R, Singh VK (2003). Anticellular and immunosuppressive properties of ethanolic extract of *Acorus calamus* rhizome. Intl Immunopharmacol 3: 53–61.

3

Achyranthes aspera Linn.

Family

Amaranthaceae

Vernacular Name

Sanskrit: Agatha; Hindi: Latjira, Chirchira; Ayurvedic: Apamarga, Chirchita; Punjabi: Kutri; Tamil: Shiru-kadaladi; Gujarati: Safed Aghedo.

Distribution [1-2]

It grows in the tropical and warmer regions of the world. The weed is also found in many countries in Asia as well as in Africa, America and Europe. It is found as weed on road sides, bare field and waste places, up to an altitude of 2100 m across India.

Traditional Uses [3-4]

Achyranthes aspera is used as diuretic in the treatment of dropsy as Ayurvedic medicine. Traditionally, the plant is useful in asthma, purgative, piles, oedema and cough. It is also useful in headache, pneumonia, renal complications, scorpion bite, snake bite and skin diseases *etc.*

Major Chemical Constituents [5-7]

Oleanolic acid, linoleic acid, palmitic acid, oleanolic acid glycoside and β-sitosterol

Plant in Natural Habitat

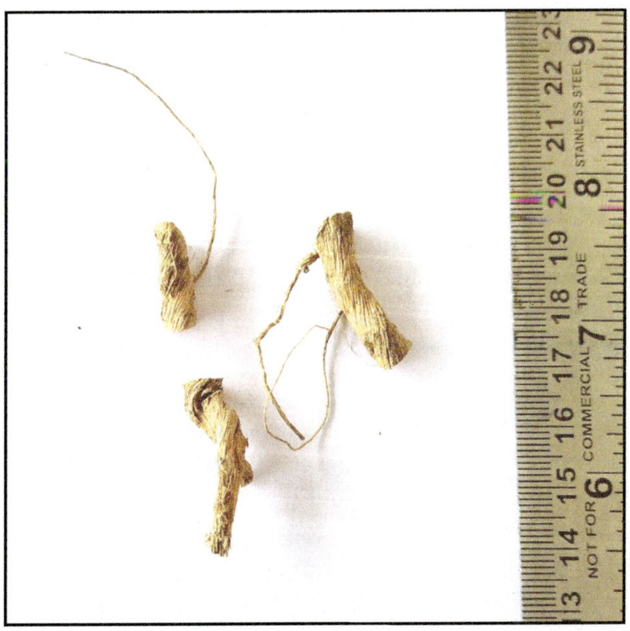

Dried Root

Pharmacological Activity [8-10]

Antioxidant, anti-inflammatory, hepatoprotective and spermicidal Activity

Plant Description

Achyranthus aspera is an erect, annual or perennial herb attain about 1-1.5 meter height with woody base. Leaves are ovate-eliptic or somewhat rounded, smooth, entire, petiolate, with prominent mid-vein and numerous lateral veins. Stem are angular, simple or branched, smooth. Inflorescence is slender spike, terminal on stem and branches. Flowers are greenish white or red in upper part of flower, bract and bracteole persistent, ending in a spine. Primary roots are cylindrical, thick, yellowish brown in color

Anatomy

Transverse section of mature root is almost circular in outline. The outer layer is made up of rectangular 7 to 10 layered cork cells, followed by 1 or 2 layered cork cambium. The cortical region is made up of 3 to 6 layered conjunctive parenchymatous cells. Vascular region occupies major area of section. Steles are well developed composed of 2-10 or more vessels associated with narrow phloem. Vessels having simple and bordered pits with helical thickening. Tracheids are also present with tapering end walls.

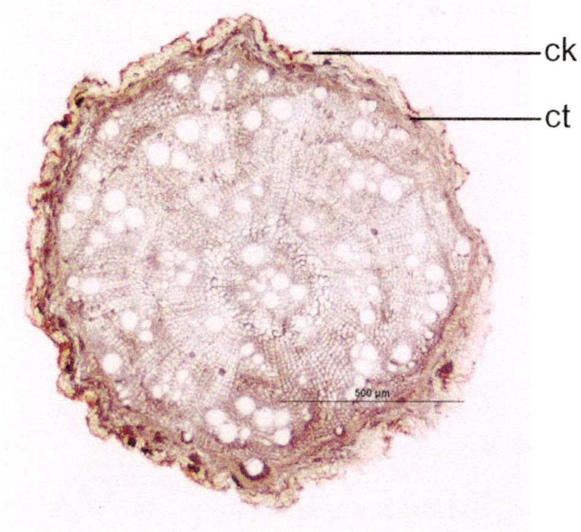

T.S. (4 x)

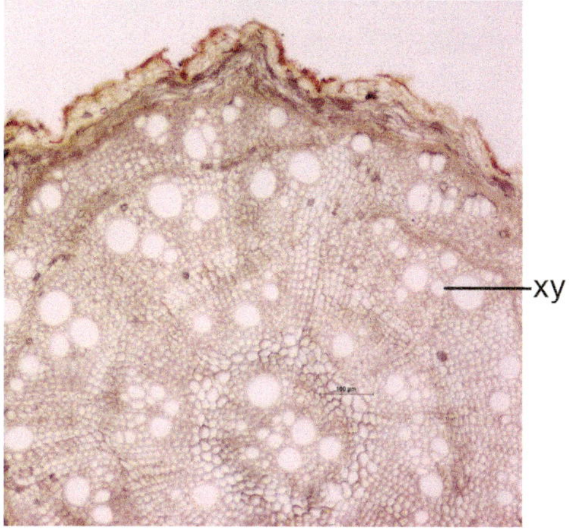

T.S. (10 x)

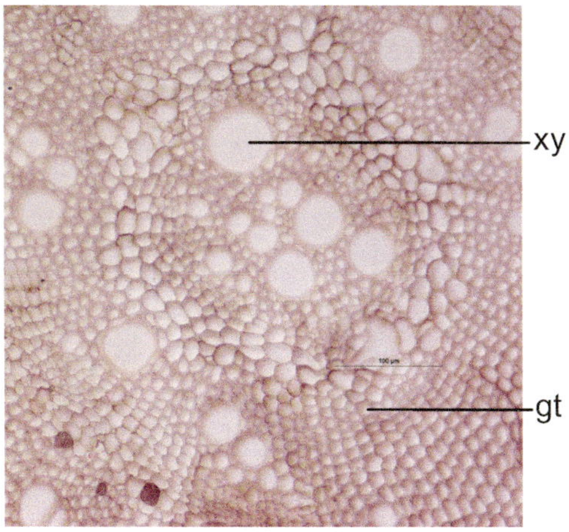

T.S. 20 x

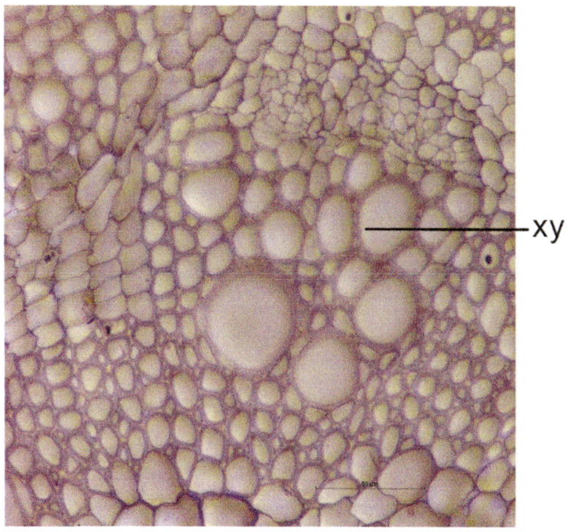

T.S. (40 x)

Transverse Section (T.S.) of *Acyranthus aspera* Root.

Abbreviation: ck- cork; ct- cortex; gt- ground tissue; xy- xylem.

HPTLC Studies

Working Solution

Accurately weighed (5g) crude powdered sample was subjected to cold extraction with methanol (50 ml), continuously shaken for 6h and then allowed to stand at room temperature (25 ± 2°C) for 18h. Extraction was repeated thrice, and the pooled filtrate was concentrate in rotatory evaporator (Buchi, Switzerland). Working solution for HPTLC study was prepared for concentration of 10 mg/ml.

Standard Solution

Oleanolic acid was prepared in fresh 1 mg/ml concentration

Sample Application

2 µl of standard solution and 10 µl of sample was applied as 6 mm band with CAMAG HPTLC applicator on activated silica gel 60 GF_{254} HPTLC plate. The chromatogram was developed on pre-saturated CAMAG HPTLC twin trough chamber for 20 min and the solvent system was allowed to run for 8 cm.

Solvent System

Toluene- ethyl acetate- formic acid (7:3:0.5 v/v).

Visualization

The chromatogram was visualized after derivatization with anisaldehyde sulphuric acid reagent and densitometric scanning was done at 530 nm.

Identification of Oleanolic Acid

The standard oleanolic acid was identified at R_f 0.67.

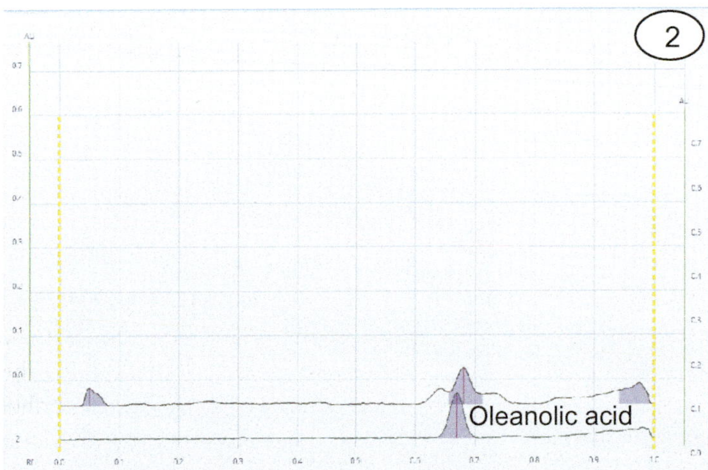

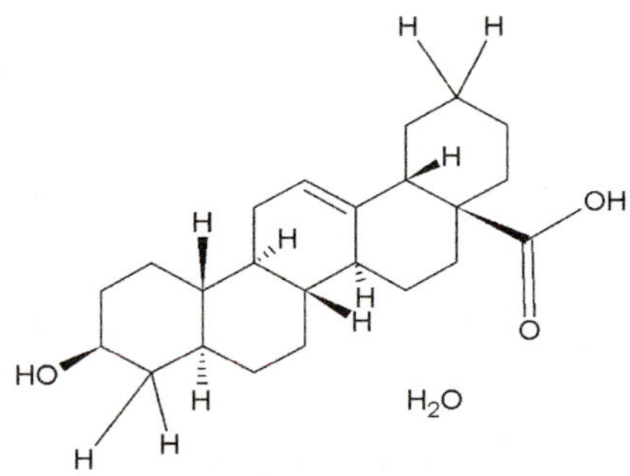

Oleanolic acid

Plate 1: HPTLC Chromatogram of *Acyranthus aspera* (S) and Standard Oleanolic Acid (R).
Plate 2: Overlay Spectra of Sample and Standard.

1. Prain D (1963). Bengal Plants, 4, Botanical Survey of India, Calcutta 646-655.
2. Shafique S, Javaid A, Bajwa R, Shafiqe S (2007). Biological control of *Achyranthes aspera* and *Xanthium strumarium* in Pakistan. Pak J Bot 39(7):2607-2610.
3. Nadkarni KM (2009). Indian Materia Medica. Bombay Popular Prakashan Vol.I 21.
4. Jain SK (1991). Dictionary of Indian folk medicine and ethnobotany. Deep Publications, New Delhi, India.
5. Kapoor and Singh (1966). A new aliphatic acid from *Achyranthes aspera* Linn. Roots. Indian J Chem 4:461.
6. Khastgir HN, Gupta PS (1958). Journal of the Indian Chemical Society 35:529-530.
7. Misra TN, Singh RS, Pandey HS, Prasad C, Singh BP (1993). Two long chain compounds from *Achyrenthes aspera*. Phytochemistry 33:221-223.
8. Bafna AR, Mishra SH (2004). Effect of methanol extract of *Achyranthes aspera* L. on rifampicin induced hepatotoxicity inrats. Ars Pharmaceutica 45(4):343-351.
9. Vijaya Kumar S, Sankar P, Varatharajan R (2009). Pharmaceutical Biology 47(10):973-975.
10. Pakrashi A, Bhattacharya N (1977). Indian Journal of Experimental Biology 15(10):856-858.

4

Aristolochia indica Linn.

Family

Aristolochiaceae

Vernacular Name

Sanskrit: Ahigandha, Arkamula, Hindi: Arkamula, kitamari; Bengali: Isarmul; Gujarati: Ruhimula; English: Indian Birthwort; Marathi: gandhaari; Tamil: Isura, Adagam.

Distribution

Aristolochia indica Linn. is globally distributed in tropical Africa, Sri Lanka, Pakistan and India. In India, it is found in northern and central India from Haryana to West Bengal, Tamil Nadu and Kerala.

Traditional Uses [1]

In Ayurveda, root is described as pungent, bitter, alexiteric and it is recommended for the treatment of snake bite. Root paste is externally applied in scabies by the ethnic people of Raipur forest division (Chattishgarh) and Nellore district (Andhra Pradesh) of India.

Major Chemical Constituents [2–3]

Oleanolic acid, aristolochic acid and aristolindiquinone.

Pharmacological Activity [4–6]

Antifertility, anti-inflammatory and anti-diarrhoeal activity.

A Fruiting Stalk

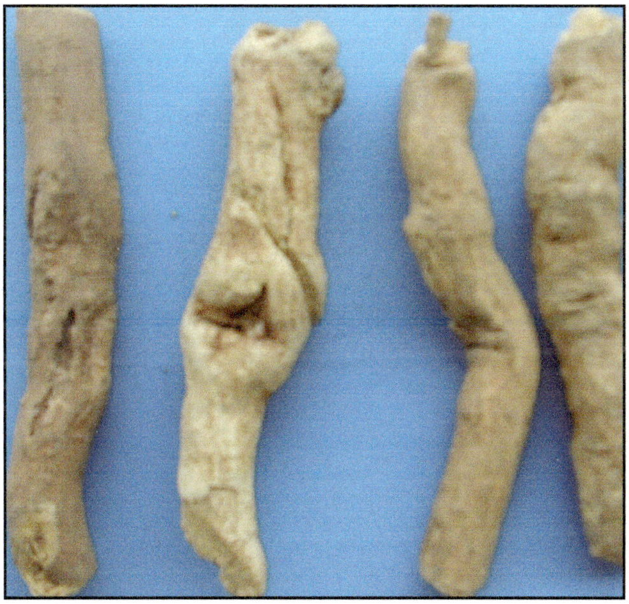

Dried Root

Plant Description

Aristolochia indica Linn. is a small perennial herb or shrub vine fleshy and soft stem with woody root. The leaves are entire, ovate-oblong, glabrous, small. Flowers are few, axillary raceme with a perianth. Stem is somewhat cylindrical, soft, fleshy and spreading prostrate. The roots are woody and friable; outer surface is brown colored, thick and inner white colored with characteristic taste and smell.

Anatomy

Transverse section of root is almost circular, shows very narrow range of cork and central wood, encircled by narrow phloem and wider cortex. Detailed structure accounted that the root shows narrow cork made of 2 to 3 row of dark brown cells, followed by wide cortex, cortical cells are round or oval shaped, consists of starch grains. Phloem narrow divided into primary and secondary phloem, cambium distinct. Lignified medullary rays in continuation with that of phloem filled with quadrangular dark brown content which may be calcium crystals and radially arranged with xylem vessels. Xylem is consists of vessels, tracheids, fibers and parenchyma. In longitudinal section xylem vessels are seen vertically running broad segmented canals with bordered pitted and thickened wall. The arrangements of medullary rays are horizontal, along with xylem vessels and fibers.

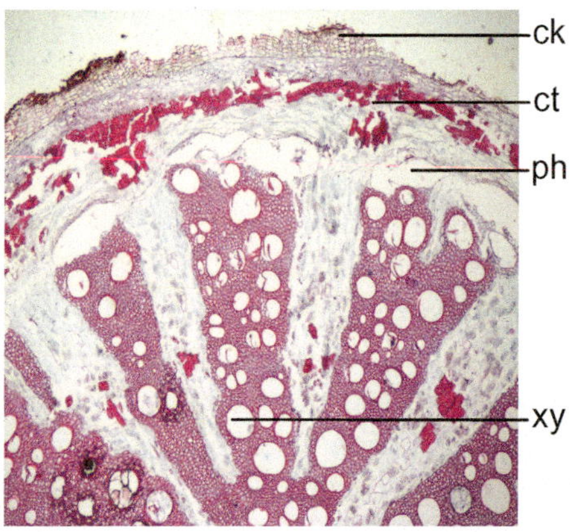

T.S. (4 x)

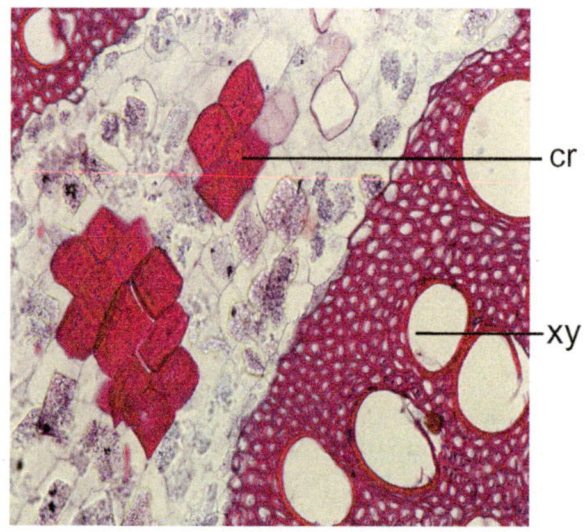

T.S. (10 x)

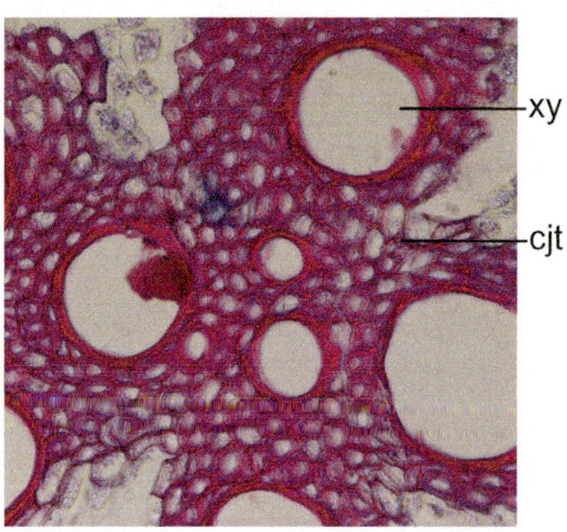

T.S. (20 x)

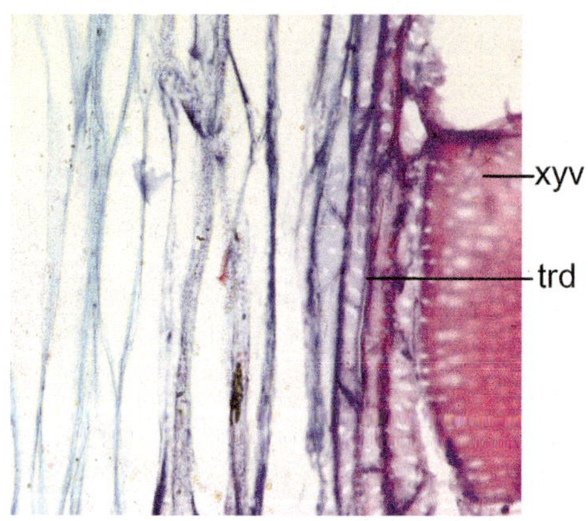

L.S. (40 x)

Transverse Section (T.S.) & Longitudinal Section (L.S.) of *Aristolochia indica* Root.

Abbreviation: ck- cork; ct- cortex; cr- crystals; cjt- conjunctive tissue; mr- medullary rays; ph- phloem; trd- tracheid; xy- xylem; xyv- xylem vessels.

HPTLC Studies

Sample Preparation

Accurately weighed (5g) crude powdered sample was subjected to cold extraction with methanol (50 ml), continuously shaken for 6h and then allowed to stand at room temperature (25 ± 2°C) for 18h. Extraction was repeated thrice, and the pooled filtrate was concentrate in rotatory evaporator (Buchi, Switzerland). Working solution for HPTLC study was prepared at a concentration of 10 mg/ml.

Standard Solution

Oleanolic acid was prepared in fresh 1 mg/ml concentration

Sample Application

2 µl of standard solution and 10 µl of sample was applied as 6 mm band with CAMAG HPTLC applicator on activated silica gel 60 GF_{254} HPTLC plate. The chromatogram was developed on pre-saturated CAMAG HPTLC twin trough chamber for 20 min and the solvent system was allowed to run for 8 cm.

Solvent System

Toluene- ethyl acetate- formic acid (7:3:0.5 v/v).

Visualization

The chromatogram was visualized after derivatization with anisaldehyde sulphuric acid reagent and densitometric scanning was done at 530 nm.

Identification of Oleanolic Acid

The standard oleanolic acid was identified at R_f 0.67.

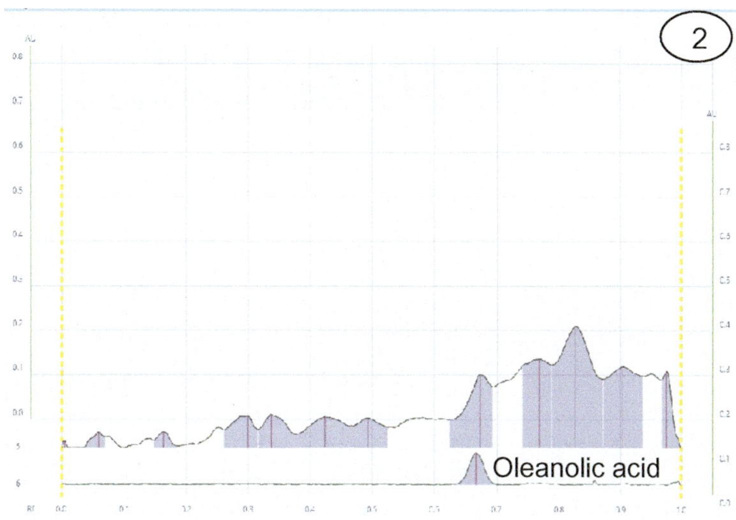

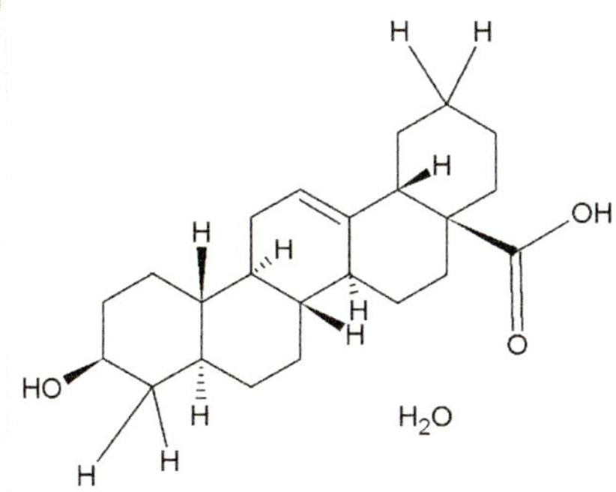

Oleanolic acid

Plate 1: HPTLC Chromatogram of *Aristolochia indica* (S) and Standard Oleanolic Acid (R), after Spraying with Anisaldehyde Sulphuric Acid Reagent.

Plate 2: Overlay Spectra of Sample and Standard.

1. Neelima M, Prasad GP, Sudarsanam G, Pratap GP, Jyothi B (2011). Ethanobotanical studies in Raipur forest division of Nellore district in Andhra Pradesh. Life Sci Leaflets 11: 333-345.
2. Mix DB, Guinaudeau H, Shamma M (1982). J Nat Prod 45: 657-666.
3. Cowe HJ, Cox PJ, Cordell GA, Che CT, Fong HH, Howie RA (1986). Structure of 2, 5-dihydroxy-3, 8-dimethyl-1, 4-naphthalenedione (aristolindiquinone). Acta Crystallographica Section C: Crystal Structure Communications 42(11):1533-35.
4. Pakrashi A, Pakrasi P (1979). Antifertility efficacy of the plant *Aristolochia indica* linn on mouse. Contraception 20(1): 49-54.
5. Rakesh D, Atul K, Pal TK (2010). Anti-inflammatory activity study of antidote *Aristolochia indica* to the venom of *Heteropneustes fossilis* in rats. J Che Pharmaceu Res 2(2): 554-62.
6. Dharmalingam SR, Madhappan R, Ramamurthy S, Chidambaram K, Srikanth MV, Shanmugham S, Kumar SK (2014). Investigation on antidiarrhoeal activity of *Aristolochia indica* Linn. root extracts in mice. African J Trad, Complementary Alter Med 11(2): 292-4.

5

Berberis aristata DC.

Family

Berberidaceae

Vernacular name

Sanskrit: Daruharidra, Kata, Suvarnavarna; Hindi: Dar-hald, Chitra; Bengali: Darhaldi; Marathi: Daruhald; English: Tree turmeric, Indian barberry; Garhwali: Kashmoi; Punjabi: Chitra, Simlu; Tamil: Usikkala.

Distribution

Berberis aristata is native to Himalayas of India and, Nepal and found at an elevation of 2000 to 3500 meters. The species is also found in Nilgiri range in Southern India.

Traditional Uses [1-3]

B. aristata (mixed with honey) is useful in the treatment of external skin lesions and ulcers. The decoction of root bark of *B. aristata* is used for washing painful eyes (externally), wounds and ulcer. In the Unani system of medicine it is used for the treatment of leprosy. The decoction of the roots is used in skin disorders and blood purification.

Major Chemical Constituents [4-5]

Berberine, berbamine, palmitine, jatrorrhizine, epiberberine.

A Fruiting Stalk

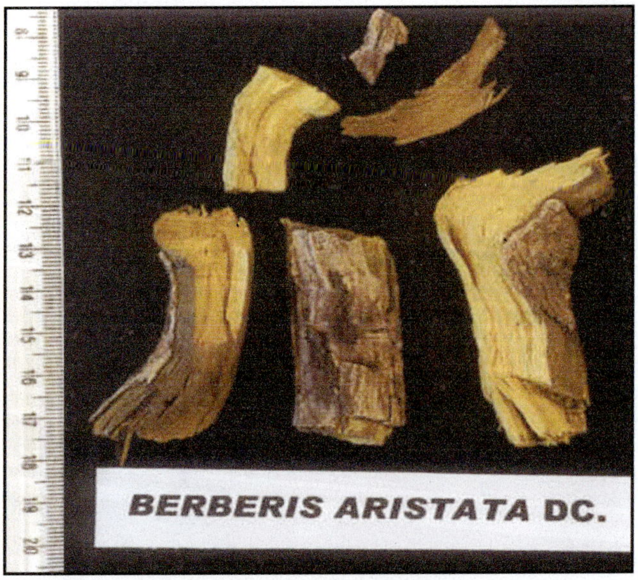

Dried Piece of Root

Pharmacological Activity [6-9]

Anti-inflammatory, anti-diabetic, cardio vascular, anti-amoebic and hepatoprotective activity.

Plant Description

Berberis aristata DC. is a large deciduous shrub usually 1.5-4 m high. Twigs are yellowish brown, erect cylindrical, smooth and internally bright yellow. Leaves are obovate or elliptic, entire or spinous-toothed, with prominent reticulate nerves. Flowers are numerous; inflorescence simple drooping raceme, bracts small, linear, acuminate; sepals imbricate, yellowish; petals 6 arranged in two whorls; stamens 6 equal, hypogynous, opposite and little shorter than the petals; ovary simple, single celled. Fruit is small berry. The roots are thick, woody, yellow, cylindrical, covered with a thin friable bark. Bark is internally yellowish, rough, fracture hard, odorless and bitter in taste.

Anatomy

Transverse section of mature root is circular in outline. The outer layer is made up of rectangular 15 to 19 layered cork cells. The cortical region is made up of 30 to 35 layered irregular parenchymatous cells; outer 4 to 6 layers are compactly arranged and contain sclereids, while the inner cortical cells are broad having intercellular spaces. Sclereids are present mostly solitary and rarely grouped. Just below the cortex 4 to 10 layered thick phloem having sieve tubes, companion cells and phloem fibers. Vascular cambium is 2 or 3 layered. The xylem is 8-10 celled wide and consists of vessels, fibers, tracheids and xylem parenchyma. The medullary rays are thick-walled, radiating, and parenchymatous. In longitudinal section vessels are seen vertically running broad segmented canals with bordered pitted and thickened wall, parenchyma are seen as narrow elongated tubes. The arrangements of medullary rays are horizontal, along with xylem vessels and fibers; tracheids are found along with the xylem vessels.

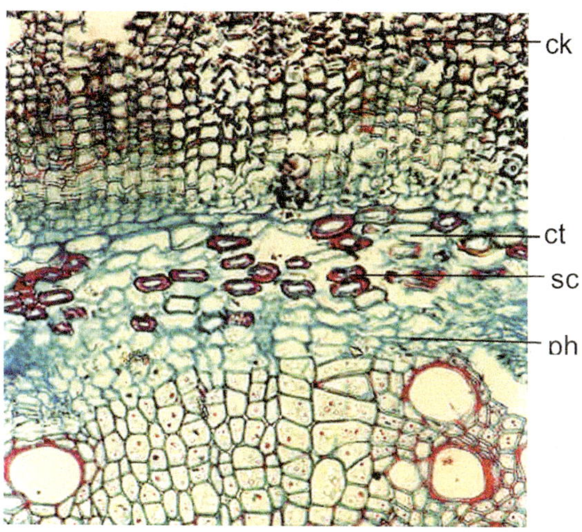

T.S. (20 x)

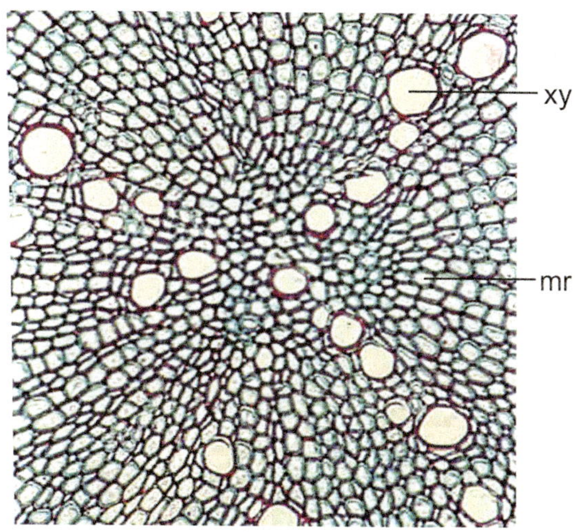

T.S. (10 x)

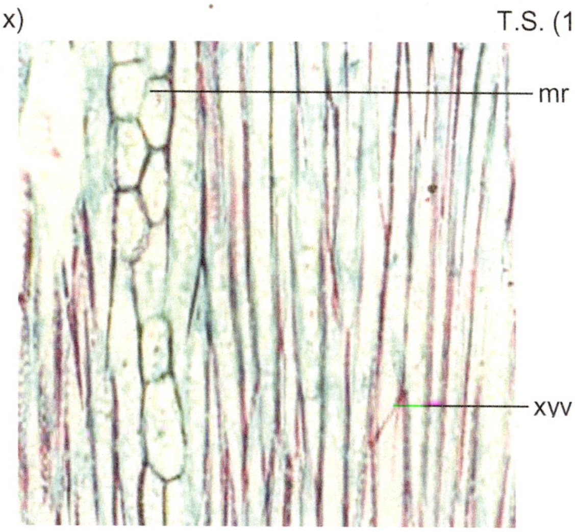

L.S. (40 x)

Transverse Section (T.S.) & Longitudinal Section (L.S) of *Berberis aristata* Root.

Abbreviation: ck- cork; ct- cortex; mr- medullary rays; ph- phloem; sc- sclereids; xy- xylem; xyv- xylem vessels.

HPTLC Studies

Sample Preparation

Accurately weight (5 g) crude powdered sample was subjected to cold extraction with methanol (50 ml), continuously shaken for 6 h and then allowed to stand at room temperature (25 °C ± 2) for 18 h. The extraction was repeated thrice and the pooled filtrate was concentrate in rotatory evaporator (Buchi, Switzerland). Working solution of strength 10 mg/ml was freshly prepared on the day of quantative analysis.

Standard Solution

Berberine (≥ 98 per cent) was freshly prepared of strength 1 mg/ml.

Sample Application

2 µl of standard solution and 10 µl of samples was applied as 6 mm band with CAMAG HPTLC applicator on activated silica gel 60 GF_{254} HPTLC plate. The chromatogram was developed on pre-saturated CAMAG HPTLC twin trough chamber for 30 min and the solvent system was allowed to run for 8 cm.

Solvent System

n-propanol- water- formic acid (9:1:0.1 v/v).

Visualization

The chromatogram was visualized at UV_{366} and densitometric scanning of berberine was done at 350 nm.

Identification of Berberine [10]

The standard berberine was identified at Rf 0.32.

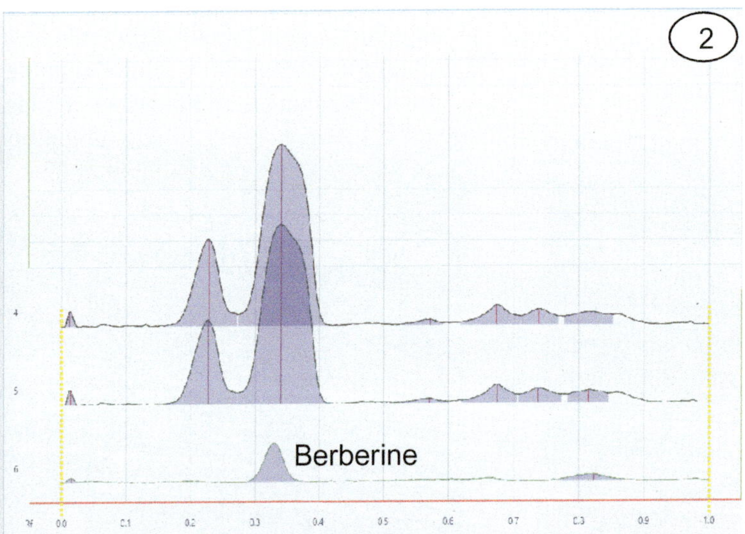

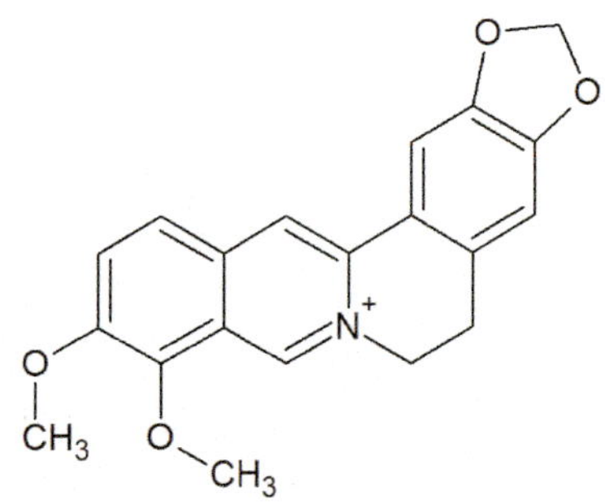

Berberine

Plate 1: HPTLC Chromatogram of *Berberis aristata* Sample (S) and Standard Berberine (R).
Plate 2: Overlay Spectra of Sample and Standard.

1. Anonymous (1988). The wealth of India, a dictionary of Indian Raw Materials, vol 2 (B). Publications and Information Directorate CSIR, New Delhi: 114-118.
2. Sharma PC, Yelne MB, Dennis TJ (2000). Database on medicinal plants used in Ayurveda. Vol. 1. New Delhi: Central Council for Research in Ayurveda and Siddha: 121.
3. Ray and Roy (1941). Folkloric uses of *Berberis aristata*. Sci and cult b13 (6).
4. Chatterjiee RP (1951). Isolation of new phytoconstituents from the plants of Berberidaceae family. J Indian chem Soc 28: 225.
5. Saied S, Batool S, Naz S (2007). Phytochemical studies of *Berberis aristata*. J basic applied sci 3(1): 1-4.
6. Akhter MH, Sabir M, Bhide NK (1977). Anti-inflammatory effect of Berberine on rats injected locally with cholera toxin. Indian J Med Res 65: 133-44.
7. Pareek A, Suthar M (2010). Antidiabetic activity of extract of *Berberis aristata* root in streptozotocin induced diabetic rats. Pharmacol online 2: 179-185.
8. Sohni YR, Kaimal P, Bhatt RM (1995). The antiamoebic effect of a crude drug formulation of herbal extracts against Entamoeba histolytica in vitro and vivo. J Ethnopharmacol 1: 43-52.
9. Yogesh HS, Chandrashekhar VM, Katti HR, Ganpaty S, Raghavendra HL, Gowda GK (2011). Antiosteoporotic activity of aqueous-methanol extract of *Berberis aristata* in ovariectomized rats. J Ethnopharmacol 134: 334-8.
10. Srivastava SK, Khatoon S, Rawat AK, Mehrotra S, Pushpangadan P. (2017). Pharmacognostic Evaluation of the Root of *Berberis aristata* DC. Natural Product Sciences, 7(4): 102-6.

6

Berberis asiatica Roxb.

Family

Berberidaceae

Vernacular Names

Sanskrit: Daruharidra, Kata, Suvarnavarna; Hindi: Dar-hald, Chitra; Bengali: Darhaldi; Marathi: Daruhald; English: Tree turmeric, Indian barberry; Garhwali: Kashmoi; Punjabi: Chitra, simlu; Tamil: Usikkala.

Distribution [1]

Berberis asiatica Roxb is the widely distributed species of the Western Himalayas and Northeast India ranging from states of Jammu and Kashmir, Himachal Pradesh, Uttarakhand, Arunachal Pradesh and Assam.

Traditional Uses [1-2]

In the Yunani medicine system *B. asiatica* Roxb plant are used to treat jaundice, eye sores, toothache, asthma and skin pigmentation, drying unhealthy ulcers, for removing inflammation and swelling.

Major Chemical Constituents [3]

Berberine, berbamine, palmitine, jatrorrhizine.

Pharmacological Activity [4-6]

Anti-oxidant, anti-bacterial, anti-microbial and hepato-protective activity.

Plant Description

An evergreen shrub, 1-2 m. high and stem cylindrical, bark rough, withered and corky; twigs are glabrous,

A Flowering Twig

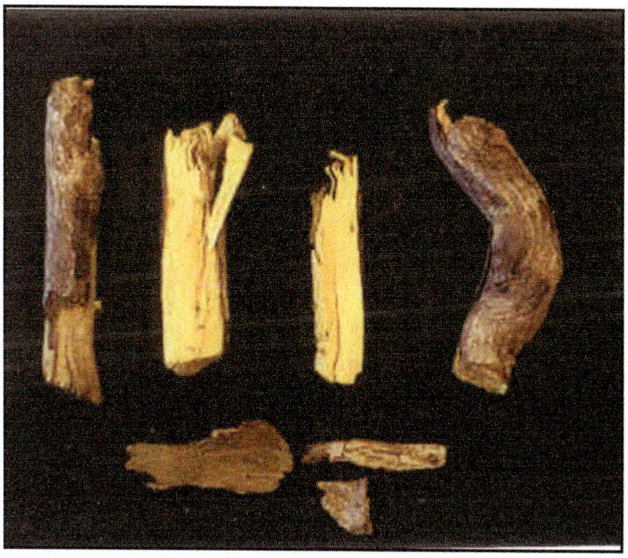

Dried Pieces of Root

pale yellowish. Leaves are oblong, elliptic or broadly obovate and dentate, sometime entire, dark green with prominent primary and secondary reticulate venation. Inflorescence is simple raceme up to 3 cm long, often with a few long-stalked flowers at the base; fruit 7-10 mm long, ovoid, blue-black with distinct style. Root hard, cylindrical, 5 cm or more in diameter, pale yellow and distinctly radiated. Root bark is 2 mm thick, brownish externally corky and friable internally smooth, closely striated and yellow in colour. Fracture very hard and taste very bitter.

Anatomy

The transverse section of root is circular in outline shows narrow range of cork and central wood, encircled by narrow phloem and wider cortex. The outermost zone is made of cork cell which are rectangular and 10 to 13 layered, consists 1 to 2 layered cork cambium. Cortex, 15 to 18 layered, narrowed due to secondary growths; filled with tannins, starch grains and sclereids. Sclereids are in groups of 2 to 10 or more and rarely solitary. Pericycle is interrupted by groups of stone cells. The xylem is polyarch, with 2 to 4 cells wide, heterogeneous, radiating medullary rays; most of the medullary rays are filled with starch gains; 1-2-layered cambiums are also present. The central zone is narrow and mostly occupied by tracheids, xylem fibers and few vessels. In longitudinal section, rays are heterogeneous and arrangements of medullary rays are horizontal, along with vessels and fibers; tracheids are found along with the xylem vessels.

HPTLC Studies

Sample Preparation

Accurately weight (5 g) crude powdered sample was subjected to cold extraction with methanol (50 ml),

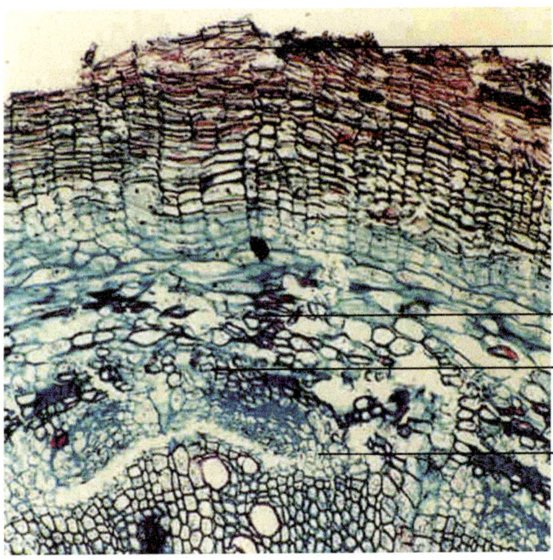

T.S. (10 x)

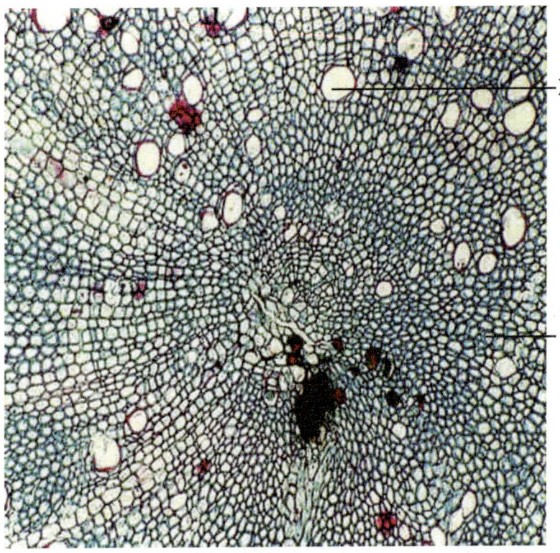

T.S. (10 x)

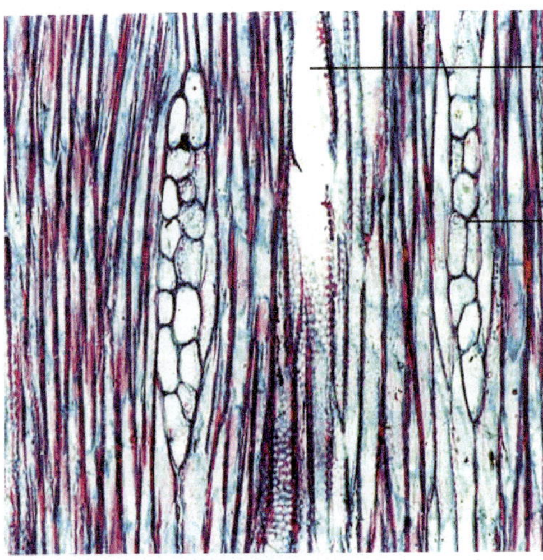

L.S. (40 x)

Transverse Section (T.S.) & Longitudinal Section (L.S) of *Berberis asiatica* Root.

Abbreviation: ck- cork; ct- cortex; mr- medullary rays; ph- phloem; sc- sclereids; xy- xylem; xyv- xylem vessels.

continuously shaken for 6 h and then allowed to stand at room temperature (25 °C ± 2) for 18 h. The extraction was repeated thrice, filtered and the pooled filtrate was concentrate in rotatory evaporator (Buchi, Switzerland). Working solution of strength 10 mg/ml was freshly prepared on the day of quantative analysis.

Standard Solution

Berberine (≥ 98 per cent) was freshly prepared of strength 1 mg/ml.

Sample Application

2 µl of standard solution and 10 µl of sample was applied as 6 mm band with CAMAG HPTLC applicator on activated silica gel 60 GF_{254} HPTLC plate. The chromatogram was developed on pre-saturated CAMAG HPTLC twin trough chamber for 30 min and the solvent system was allowed to run for 8 cm.

Solvent System

n-propanol: water: formic acid (9:1:0.1 v/v).

Visualization

The chromatogram was visualized at UV_{366} and densitometric scanning of berberine was done at 350 nm.

Identification of Berberine[7]

The standard berberine was identified at R_f 0.32.

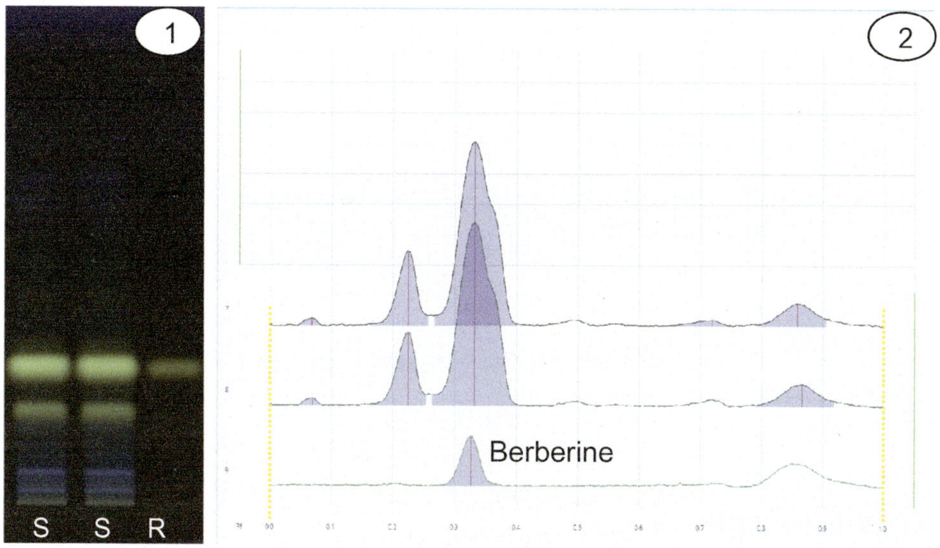

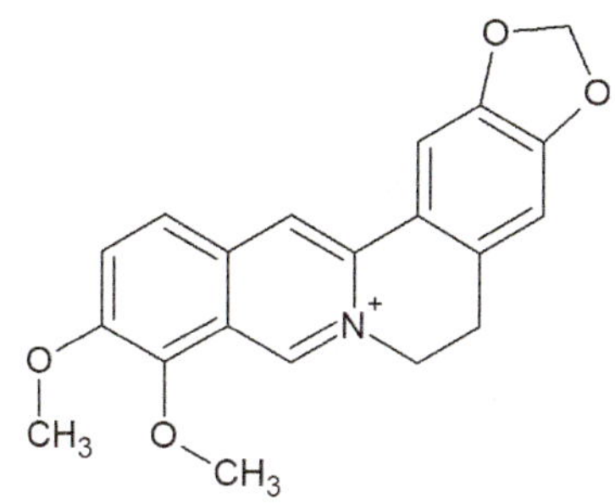

Berberine

Plate 1: HPTLC Chromatogram of *Berberis asiatica* Sample (S) and Standard Berberine (R).
Plate 2: Overlay Spectra of Sample and Standard.

1. Kirtikar KR, Basu BD (1998). Indian medicinal plants, vol 1. CSIR publication, India.
2. Anonymous (1988). The wealth of India, a dictionary of Indian Raw Materials, vol 2 (B). Publications and Information Directorate CSIR, New Delhi: 114-118.
3. Chatterjiee RP (1951). Isolation of new phytoconstituents from the plants of Berberidaceae family. J Indian chem Soc 28: 225.
4. Bhandari DK, Nath G, Ray AB, Tewari PV (2000). Antimicrobial activity of crude extract from *Berberis asiatica* stem bark. Pharm Biol 38: 254-257.
5. Hashmi K, Hafiz A (1986). In vivo antibacterial activity of *Berberis asiatica*. J Pakistan Med A 36: 5-7.
6. Tiwari BK, Khosa RL (2010). Evaluation of the heptoprotective and antioxidant effect of *Berberis asiatica* against experimentally induced liver injury in rats. Int J Pharm Pharm Sci 2(1): 92-97.
7. Srivastava SK, Singh Rawat AK, Mehrotra S. (2004). Pharmacognostic evaluation of the root of *Berberis asiatica*. Pharmaceutical biology. Jan 1; 42(6): 467-73.

7

Berberis chitria Lindl.

Family

Berberidaceae

Vernacular name [1-2]

Sanskrit: Kata, Suvarnavarna; Hindi: Dar-hald, Chitra; Bengali: Darhaldi; English: Tree turmeric; Garhwali: Kashmoi; Punjabi: Chitra, Simlu.

Distribution [1-2]

B. chitria is widely distributed in Himalayas of India and Nepal at an altitude range of 2000 to 3500 meters.

Traditional Uses [3]

B. chitria is one of the components of traditional medicine "rasont" or rasanjan". It has been used since older time as home remedies for treatment of conjunctivitis, ulcers, skin diseases, jaundice and in treatment of enlarged liver and, spleen.

Major Chemical Constituents [4]

Berberine, Palmitine, jatrorrhizine

Pharmacological Activity [5-7]

Anti-microbial, anti-fertility and cardio vascular activity

A Flowering Twig

Dried Pieces of Root

Plant Description

An evergreen shrub 1.5-2.5 m high, stem rough brown, spiny; leaves are obovate lanceolate, reticulate, slightly dentate edges; inflorescence is corymbose raceme type, flowers in groups of 3-5; bract 2-3 mm long; outer sepals unequal and inward; petals clawed; stamens are connective capitates; berries are dark brown, ovoid or ellipsoidal. Root is cylindrical, hard, branched and gradually tapering; the outer surface light brown and smooth. Root is bright yellow, distinctly radiated with narrow medullary rays. The bark is up to 5 mm thick, externally fissured and internally smooth.

Anatomy

Transverse section is almost circular and regular in outline. The outer layer is cork rectangular 6 to 11 layered cork cells, followed by 2 or 3 layered cork cambium. The cortical region is made up of 15 to 22 layered irregular parenchymatous cells; outer 6 to 8 layers are contain sclereids, while the inner cortical cells are broad having intercellular spaces. Sclereids are present mostly solitary or in a group of 2 to 5. Pericycle is discontinued by scelerenchymatous cells patches. Just below the cortex 3 to 6 or more layered phloem are present having sieve tube, companion cell, fibers and parenchyma. Cambium is single layered and medullary rays thin walled, heterogenous, parenchymatous, contains starch grains and alkaloid. The xylem vessels are solitary or in group of 2-3 and center are occupied by vessels and tracheids. In longitudinal section vessels are seen vertically running broad segmented canals with bordered pitted and thickened wall. The arrangements of medullary rays are horizontal, along with xylem vessels and fibers; tracheids are found along with the xylem vessels.

HPTLC Studies

Sample Preparation

Accurately weight (5g) crude powdered sample was subjected to cold extraction with methanol (50 ml), continuously shaken for 6h and then allowed to stand at room temperature (25 °C ± 2) for 18h. Extraction was repeated thrice, filtered and the pooled filtrate was concentrate in rotatory evaporator (Buchi, Switzerland). Working solution of strength 10 mg/ml was freshly prepared on the day of quantative analysis.

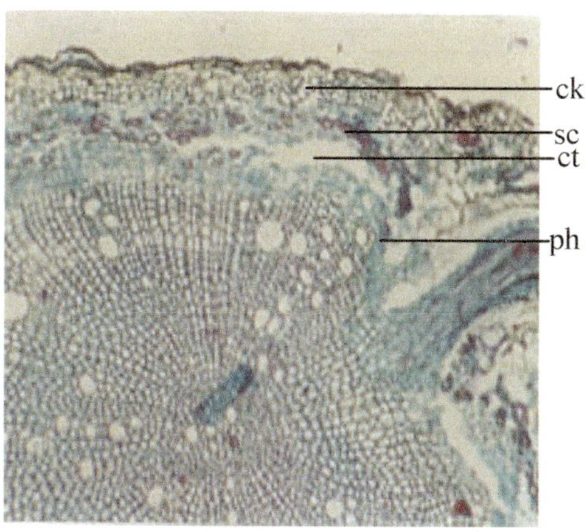

T.S. (20 x)

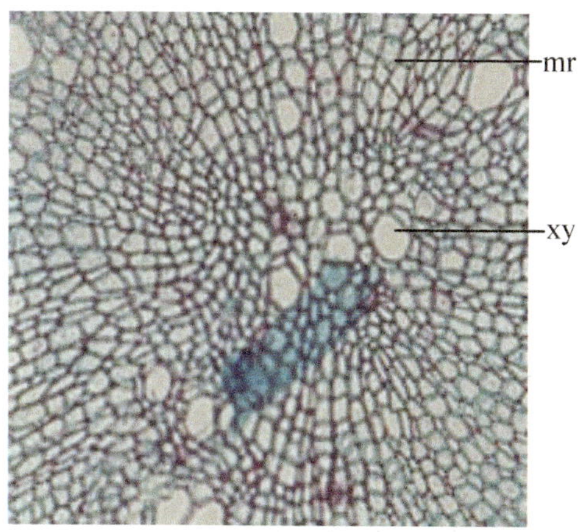

T.S. (20 x)

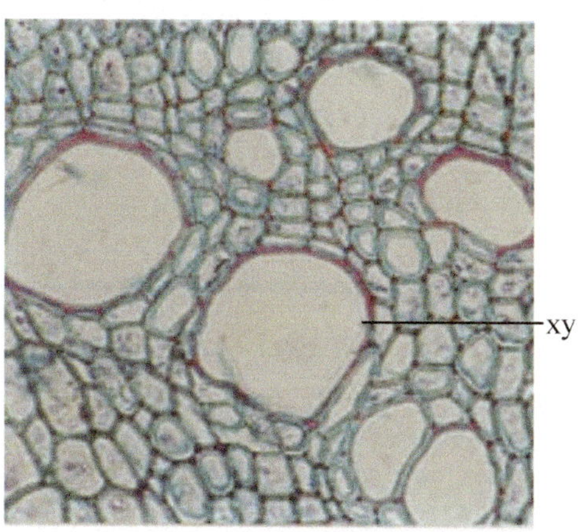

T.S. (40 x)

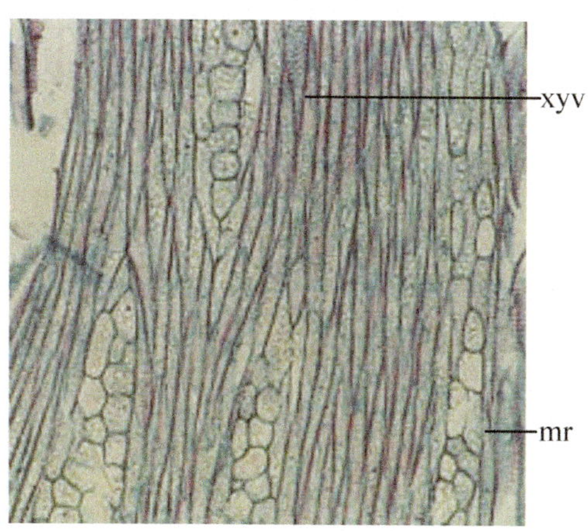

L.S. (40 x)

Transverse Section (T.S.) & Longitudinal Section (L.S.) of *Berberis chitria* Root.

Abbreviation: ck- cork; ct- cortex; mr- medullary rays; ph- phloem; sc- sclereids; xy- xylem; xyv- xylem vessels.

Standard Solution

Berberine (≥ 98 per cent) was freshly prepared of strength 1 mg/ml.

Sample Application

2 µl of standard solution and 10 µl of sample was implemented by HPTLC applicator on HPTLC preset silica gel 60 GF254 plate. The 30 minute saturation was given, and the plate was developed in saturated chamber for up to 8 centimeters.

Solvent System

n-propanol: water: formic acid (9:1:0.1 v/v).

Visualization

The chromatogram was visualized at UV_{366} and densitometric scanning of berberine was done at 350 nm.

Identification of Berberine[8]

The standard berberine was identified at R_f 0.32.

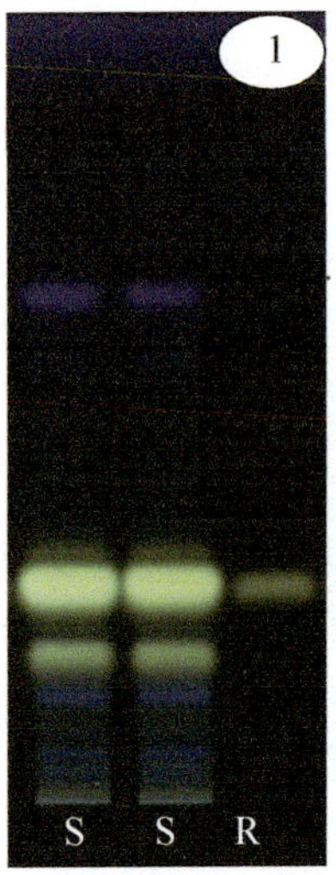

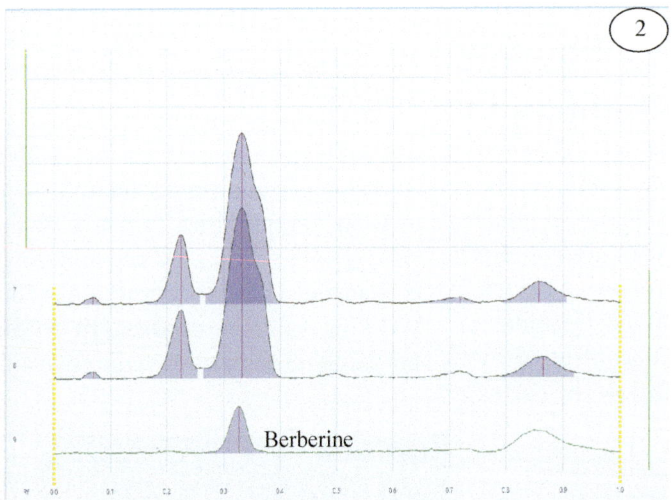

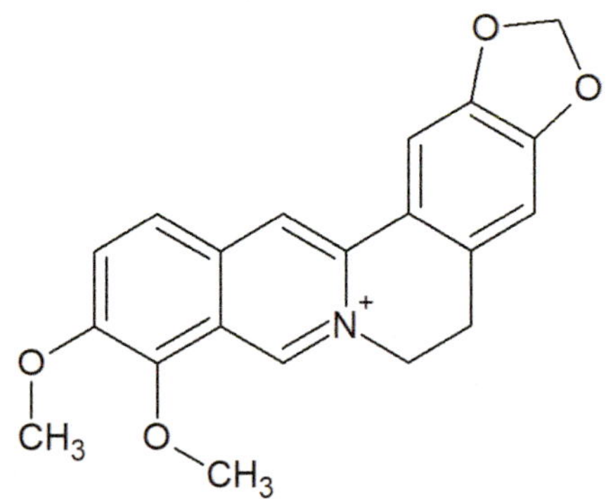

Berberine

Plate 1: HPTLC Chromatogram of *Berberis chitria* Sample (S) and Standard Berberine (R).

Plate 2: Overlay Spectra of Sample and Standard.

1. Anonymous (1988). The wealth of India, a dictionary of Indian Raw Materials, vol 2 (B). Publications and Information Directorate CSIR, New Delhi 114–118.
2. Anonymous (1985). The wealth of India, vol 1. Publications and Information Directorate CSIR, New Delhi 176–179.
3. Rajasekaran A, Kumar N (2009). Rasont-a traditional crude drug prepared from *Berberis* sp. and its uses. Indian J Tradit Know 8(4): 562-563.
4. Hussain FA, Shoeb A (1958). Isoquinoline derived alkaloids from *B. chitria*. Phytochemistry 24: 633.
5. Dobhal MP, Negi GS, Joshi BC (1988). Chemical investigation and biological activity of *B. chitria* Han. Himalayan Chem Pharm Bull 5: 41-6.
6. Gupta RS, Dixit VP (1989). Testicular cell population dynamics following palmitine hydroxide treatment in mole dogs. J Ethnopharmacol 25: 151-7.
7. Xiong C, Fang D (1989). Effect of jatrorrhizine on isolated guinea pig atria. Zhongguo Yaolixue Yu Dulixue Zazhi 3: 255-8.
8. Srivastava SK, Rawat AK, Srivastava M, Mehrotra S. (2006). Pharmacognostic Evaluation of the Roots of *Berberis chitria* Lindl. Natural Product Sciences. 12(1): 19-23.

8

Berberis lycium Royle.

Family

Berberidaceae

Vernacular names

Sanskrit: Daruharidra, Kata, Suvarnavarna; Hindi: Dar-Hald, Chitra; Bengali: Darhaldi; Marathi: Daruhald; English: Tree turmeric, Indian barberry; Garhwali: Kashmoi; Tamil: Usikkala.

Distribution [1-3]

Berberis lycium is native to Nepal and also distributed in different parts of the World. It occurs in sub-tropical and temperate regions of India, stretching from Kashmir to Uttarakhand in Northern Western Himalayas.

Traditional Uses [4-5]

B. lycium Royle has been used in folk medicine against diarrhea, intestinal colic, jaundice. In Lahul province of Himanchal Pradesh (India), *B. lycium* and *B. pachyacantha* fruit is medicinally used for stomachache. Roots are used to yield a yellow dye for textiles. In Rajouri district of Jammu (India) fruits of *B. lycium* are used as coagulant and branches are made into broom for removing husk from grains.

A Fruiting Twig

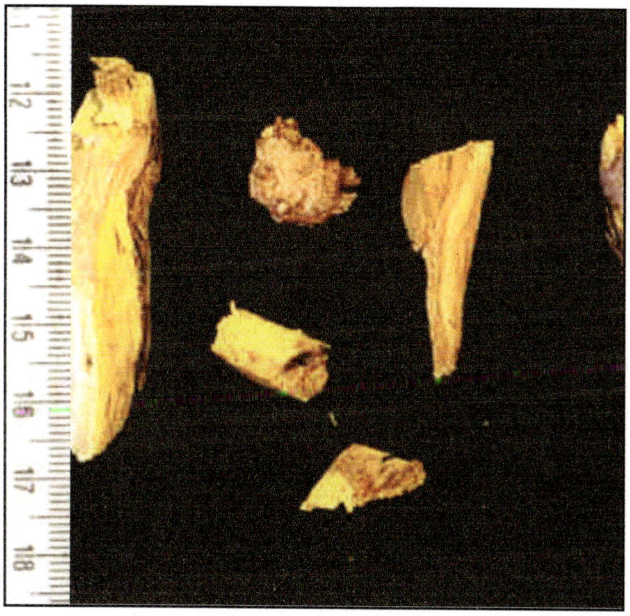

Dried Pieces of Root

Major Chemical Constituents [6]

Berberine, palmitine, jatrorrhizine.

Pharmacological Activity [7–10]

Hepatoprotective, wound healing, anti-bacterial and anti-diabetic activity.

Plant Description

Berberis aristata DC. is a large deciduous shrub usually 1.5-3.8 m high; twigs yellowish brown, erect cylindrical, smooth and bright yellow. Leaves are obovate or elliptic, entire or spinous-toothed, with prominent reticulate nerves. flowers numerous; inflorescence simple drooping raceme, bracts small, linear, acuminate; sepals 8 or 9, imbricate, yellowish; petals 6, in two whorls, fruit a small berry, ovoid, blue black with the persistent style and stigma. The roots are thick, woody, yellow, cylindrical, covered with a thin friable bark. Bark is internally yellowish, rough, fracture hard, odorless and bitter in taste.

Anatomy

Transverse section of mature root is almost circular in outline. The outer layer is made up of rectangular 15 to 19 layered cork cells, consists 2 or 3 layered cork cambium. The cortical region is made up of 30 to 35 layered irregular parenchymatous cells; outer 4 to 6 layers are compactly arranged and contain sclereids, while the inner cortical cells are broad having intercellular spaces, larger towards central region. Sclereids are observed in cortical region, mostly solitary or in a group of 3 to 8. Just below the cortex 4 to 10 layered thick phloem having sieve tubes, companion cells and phloem fibers are present. Vascular cambium is 2 or 3 layered. The xylem is 8-10 celled wide and consists of vessels, fibers, tracheids and xylem parenchyma. The medullary rays are thick-walled, radiating, and parenchymatous. In longitudinal section vessels are seen vertically running broad segmented canals with bordered pitted and thickened wall. The arrangements of medullary rays are horizontal, along with xylem vessels and fibers; tracheids are found along with the xylem vessels.

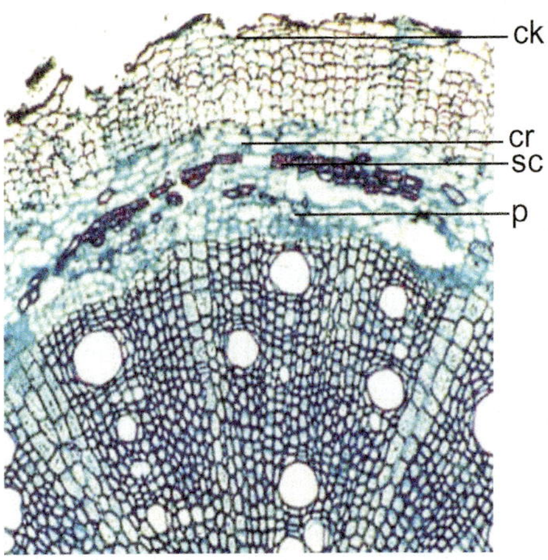

T.S. (20 x)

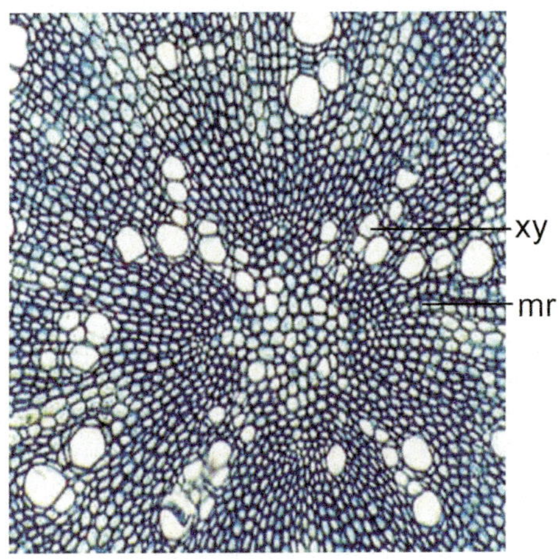

T.S. (20 x)

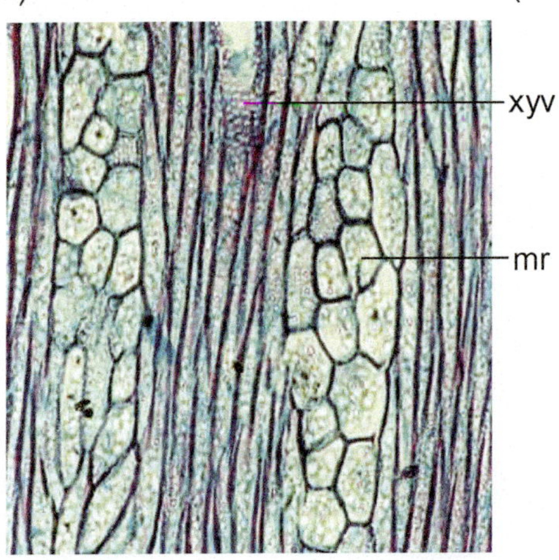

L.S. (40 x)

Transverse Section (T.S.) & Longitudinal Section (L.S) of *Berberis lycium* Root.

Abbreviation: ck- cork; ct- cortex; mr- medullary rays; ph- phloem; sc- sclereids; xy- xylem; xyv- xylem vessels.

HPTLC Studies

Sample Solution

Accurately weight (5 g) crude powdered sample was subjected to cold extraction with methanol (50 ml), continuously shaken for 6 h and then allowed to stand at room temperature (25 °C ± 2) for 18 h. The extraction was repeated thrice, filtered and the pooled filtrate was concentrate in rotatory evaporator (Buchi, Switzerland). Working solution of strength 10 mg/ml was freshly prepared on the day of quantative analysis.

Standard Solution

Berberine (≥ 98 per cent) was freshly prepared of strength 1 mg/ml.

Sample Application

2 µl of standard solution and 10 µl of sample was applied as 6 mm band with CAMAG HPTLC applicator on activated silica gel 60 GF_{254} HPTLC plate. The chromatogram was developed on pre-saturated CAMAG HPTLC twin trough chamber for 30 min and the solvent system was allowed to run for 8 cm.

Solvent System

n-propanol: water: formic acid (9:1:0.1 v/v).

Visualization

The chromatogram was visualized at UV_{366} and densitometric scanning of berberine was done at 350 nm.

Identification of Berberine[11]

The standard berberine was identified at R_f 0.32.

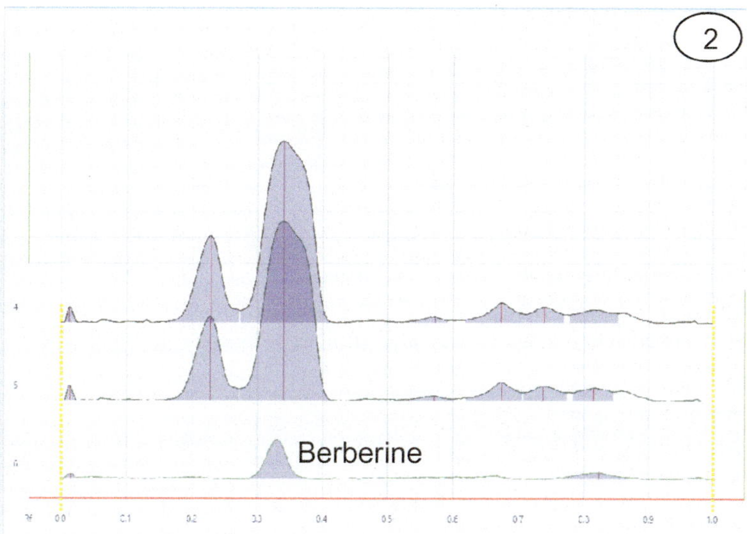

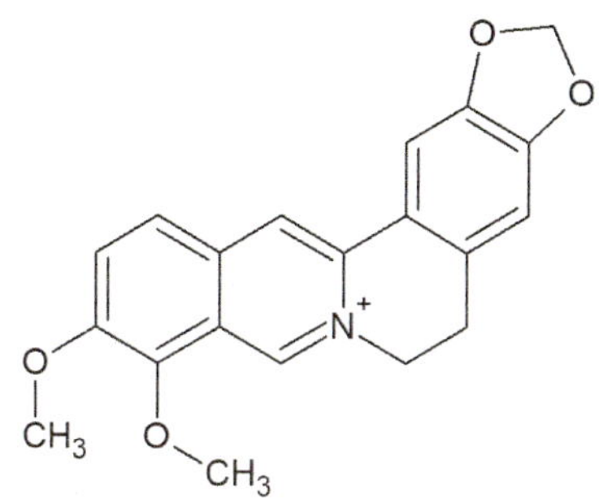

Berberine

Plate 1: HPTLC Chromatogram of *Berberis lycium* Sample (S) and Standard Berberine (R).
Plate 2: Overlay Spectra of Sample and Standard.

1. Sharma R (2003). Medicinal Plants of India; An Encyclopaedia: 33.
2. Kaur C, Miani SB (2001). Fruits and vegetables health foods for new millennium. Indian Hort 45(4): 29-32.
3. Baquar SR (1989). Medicinal and Poisonous plants of Pakistan, (Karachi, Pakistan: Printas) 55.
4. Koetz N, Walter J (1979). Notes on the ethnobotany of Lahul, a province of the Punjab. Quart J Crude Drug Res: 17: 1-56.
5. Virjee GH, Kachroo P, Bhat GM. Ethnobotanical studies of the rural areas in District Rajouri (Jammu). J Eco Taxo Bot. 1984;5:831-8
6. Bhardwaj D, Kaushik N (2012). Phytochemical and pharmacological studies in genus *Berberis*. Phytochemistry reviews 11: 523-542.
7. Girish C, Koner BC, Jayanthi S, Rao KR, Rajesh B (2009). Hepatoprotective activity of six polyherbal formulations in paracetamol induced liver toxicity in mice Pradhan. Indian J Med Res 129: 569.
8. Asif A, Kakub G, Mehmood S, Khunum R, Gulfraz M (2007). Wound Healing Activity of Root Extracts of *Berberis lycium* Royle in Rats. Phytother Res 21: 589-591.
9. Irshad AH, Pervaiz AH, Abrar YB, Fahelboum I, Awen BZ ()2013. Antibacterial activity of Berberis lycium root extract. Trakia J Sci 11(1): 89.
10. Nangyal H (2014). Advances in Biomedicine and Pharmacy (An International journal of Biomedicine, Natural products and Pharmacy) 1(1): 11-14.
11. Srivastava SK, Rawat AK, Mehrotra S. (2010). Pharmacognostic evaluation of the roots of *Berberis lycium* royle. Oriental Pharmacy and Experimental Medicine. 10(3): 184-90.

9

Berberis tinctoria Lesch.

Family

Berberidaceae

Vernacular name

Sanskrit: Daruharidra; Hindi: Chitra; Bengali: Darhaldi; Marathi: Daruhald; English: Tree turmeric, Nilgiri barberry; Garhwali: Kashmoi; Tamil: Usikkala.

Distribution

Berberis tinctoria is found in the inner Shola forest of the Nilgiri and Palani hills, at an altitude of 1,800 m.

Traditional Uses [1–4]

Berberis tinctoria root is used for treating a variety of ailment such as rheumatism, jaundice, fever, skin disease, malarial fever, eye and ear diseases. Todas of Nedimand tribes (Nilgiris, India) grind the roots of *B. tinctoria* with water and use it for stomachache, especially in the treatment of worms.

Major Chemical Constituents [5]

Berberine, berbericine, palmitine, jatrorrhizine.

Pharmacological Activity [6–7]

Hepatoprotective, anti-oxidant and anti-viral activity.

A Vegetative Plant

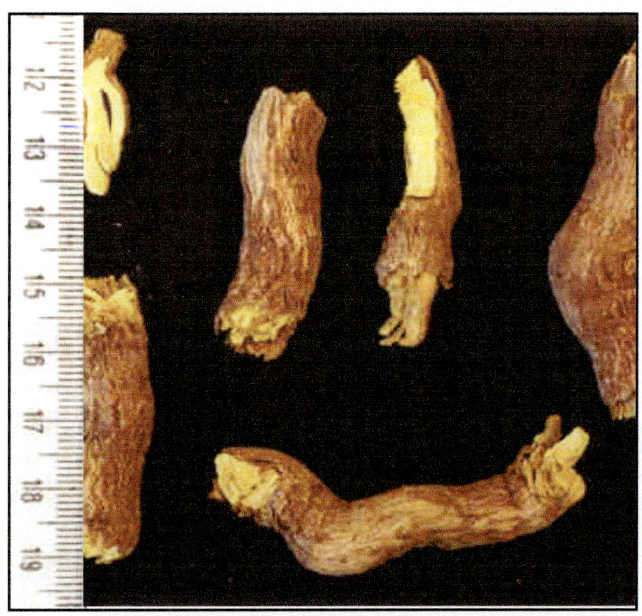

Dried Roots

Plant Description

Berberis tinctoria Lesch. is a large deciduous shrub usually 1.5-3.8 m high. Stem is erect cylindrical, smooth, bright yellow, scandent branches bearing numerous slender leafy twigs. Leaves are obovate, entire or with few spiny teeth, glaberous with prominent reticulate nerves. Flowers are numerous, inflorescence simple drooping raceme, bracts small, linear, acuminate. Berry sausage-shaped when young, eventually top-shaped, purplish red, turning to a dark blue with glaucous bloom, with dry style and large round stigma still attached. Roots are thick, woody, yellow, cylindrical, smooth surface covered with a thin friable bark. The main feature distinguishing it from other species are a smooth surface of grayish white colour and a deep yellow wood.

Anatomy

Transverse section of mature root is circular in outline shows narrow range of cork and central wood, encircled by narrow phloem and wider cortex. The outer layer is made up of rectangular 5 to 9 layered cork cells. The cortical region is made up of 16 to 20 layered irregular parenchymatous cells; outer 3 to 4 layers are compactly arranged and contain sclereids, while the inner cortical cells are broad having intercellular spaces, larger towards central region. Sclereids are mostly in groups and sometime solitary. Just below the cortex 4 to 8 layered thick phloem having sieve tubes, companion cells and phloem fibers are present. Cambium is 2 or 3 layered. The xylem is 8-10 celled wide and consists of vessels, fibers, tracheids and xylem parenchyma. Vessels are mostly solitary or in groups of 2 or 3. The medullary rays are thick-walled, radiating, and parenchymatous. In longitudinal section vessels are seen vertically running broad segmented canals with bordered pitted and thickened wall. The arrangements of medullary rays are heterogeneous, 2-5 celled broad, pitted and filled with starch grains and alkaloidal content.

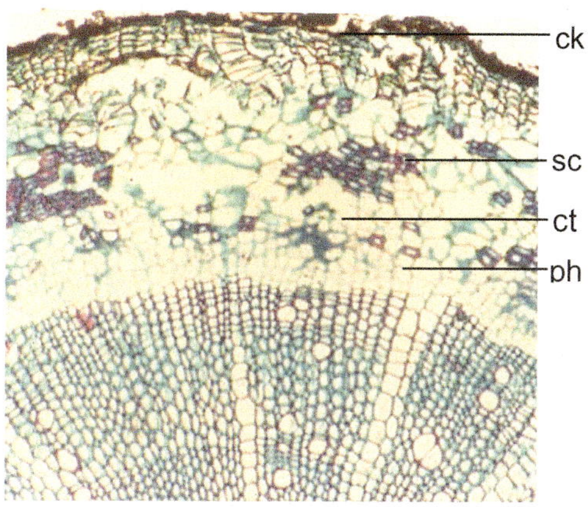

T.S. (10 x)

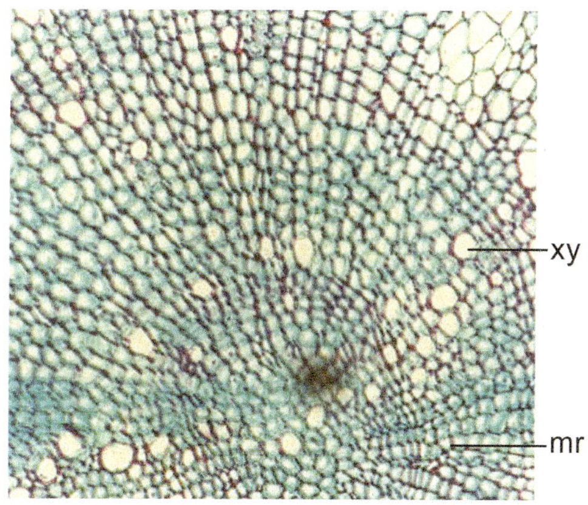

T.S. (10 x)

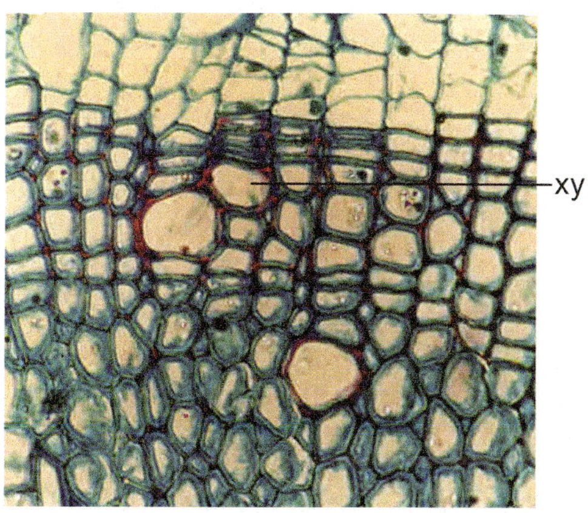

T.S. (20 x)

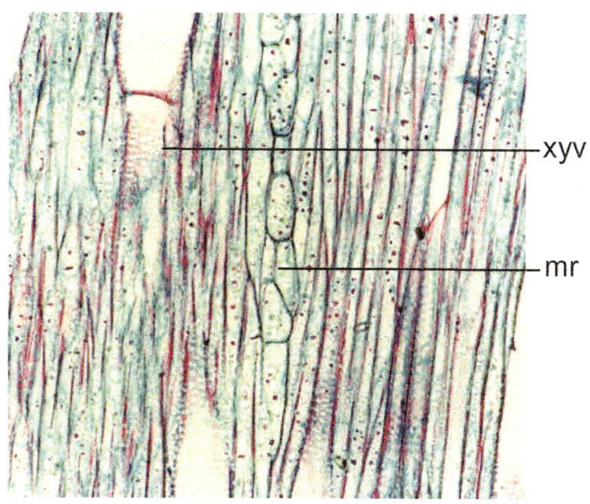

L.S. (20 x)

Transverse Section (T.S.) & Longitudinal Section (L.S) of *Berberis tinctoria* Root.

Abbreviation: ck- cork; ct- cortex; mr- medullary rays; ph- phloem; sc- sclereids; xy- xylem; xyv- xylem vessels.

HPTLC Studies

Sample Solution

Accurately weight (5 g) crude powdered sample was subjected to cold extraction with methanol (50 ml), continuously shaken for 6 h and then allowed to stand at room temperature (25 °C ± 2) for 18 h. The extraction was repeated thrice and the pooled filtrate was concentrate in rotatory evaporator (Buchi, Switzerland). Working solution of 10 mg/ml was freshly prepared on the day of quantative analysis.

Standard Solution

Berberine (≥ 98 per cent) was freshly prepared of strength 1 mg/ml.

Sample Application

2 µl of standard solution and 10 µl of sample was applied as 6 mm band with CAMAG HPTLC applicator on activated silica gel 60 GF_{254} HPTLC plate. The chromatogram was developed on pre-saturated CAMAG HPTLC twin trough chamber for 30 min and the solvent system was allowed to run for 8 cm.

Solvent System

n-propanol: water: formic acid (9:1:0.1 v/v).

Visualization

The chromatogram was visualized at UV_{366} and densitometric scanning of berberine was done at 350 nm.

Identification of Berberine[8]

The standard berberine was identified at R_f 0.32.

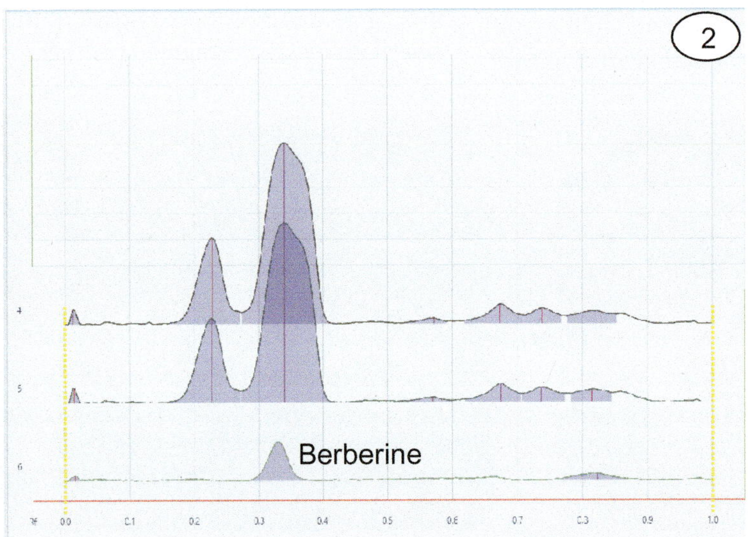

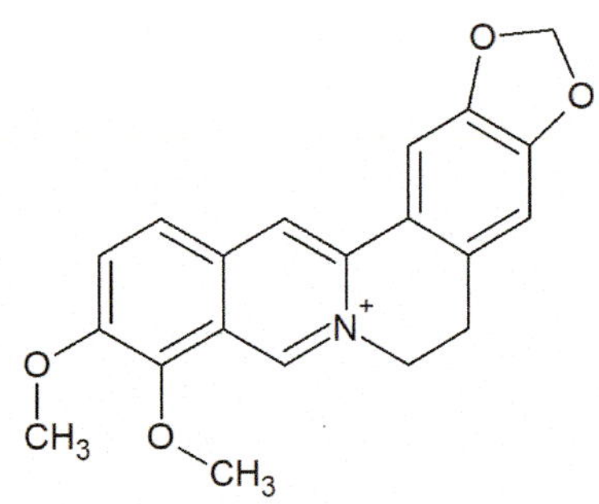

Berberine

Plate 1: HPTLC Chromatogram of *Berberis tinctoria* Sample (S) and Standard Berberine (R).
Plate 2: Overlay Spectra of Sample and Standard.

1. Anonymous (1988).The wealth of India *Berberis* Linn. (Berberidaceae). Ed Ambastha Sp Publication and Information Directorate, CSIR, New Delhi, India 2B: 114-118.
2. Watt G (1883). Economic product of India V, The superintendent of Government Printing, India.
3. Chopra RN, Chopra TC, Handa KL, Kapoor LD (1958). Chopra's Indigenous drugs of India. UN Dhar and sons Pvt Ltd: 284.
4. Abraham Z (1981). Ethnobotany of the Todas, the Kotas and the Rules of the Nilgiris. In: Jain SK (ed): Glimpses of Indian ethnobotany. New Delhi: Oxford and IBH Publ Co: 308-20.
5. Bhardwaj D, Kaushik N (2012). Phytochemical and pharmacological studies in genus *Berberis*. Phytochemistry reviews 11: 523-542.
6. Vijayan P, Raghu C, Ashok G, Dhanaraj SA, Suresh B (2004). Antiviral activity of medicinal plants of Nilgiris. Indian J med res120: 24-9.
7. Murugesh KS, Yeligar VC, Maiti BC, Maity TK (2005). Hepatoprotective and antioxidant role of *Berberis tinctoria* Lesch Leaves on Paracetamol Induced Hepatic Damage in Rats. Iranian J Pharm Ther 4(1): 64-69.
8. Rawat AK, Srivastava SK. (2007). Pharmacognostic evaluation of the roots of *Berberis tinctoria* Lesch. Natural Product Sciences. 13(1): 27-32.

10

Berberis umbellata Wall. ex G. Don

Family

Berberidaceae

Distribution [1]

Berberis umbellata is found in the Himalayan range at an altitude of 2300-3500 meter in Jammu and Kashmir, Nilgiri hills (India) and from Kumaon to Bhutan.

Major Chemical Constituents [1]

Berberine and palmitine

Pharmacological Activity [1]

Anti-inflammatory, anti-diabetic and hepatoprotective activity

Plant Description

Berberis umbellata Wall. ex G. Don is a large deciduous shrub with 0.6-1.2 meter in height; twigs yellowish brown, erect cylindrical, smooth,bright yellow. Leaves are obovate or elliptic, glabrous, entire or spinous-toothed, with prominent reticulate nerves. Flowers are numerous; inflorescence simple drooping raceme, bracts small, linear, acuminate; sepals imbricate, yellowish; fruit a small berry. The roots are thick, woody, yellow, cylindrical, covered with a thin friable bark. Bark is internally yellowish, rough, fracture hard, odorless and bitter in taste.

Anatomy

Transverse section of mature root is circular in outline. The outer layer is made up of 4-8 rows of

A Vegetative Plant

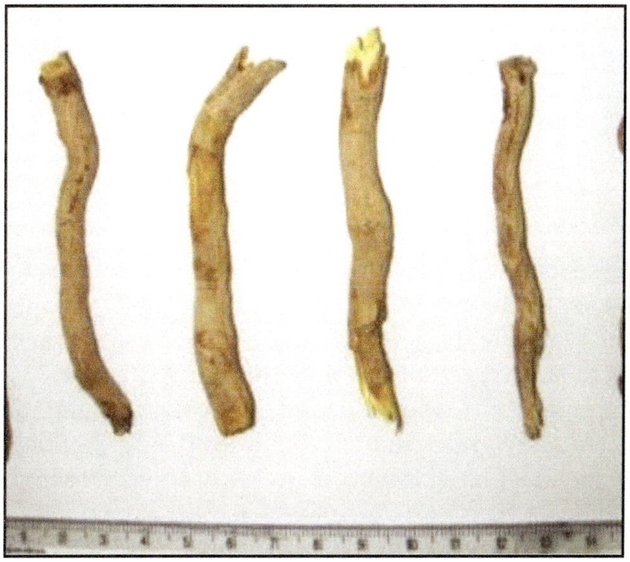

Dried Root

compactly arranged thick walled suberized rectangular cork cells. The cortical region is made up of 10 to 15 layered irregular parenchymatous cells; outer 4 to 6 layers are compactly arranged and contain sclereids, while the inner cortical cells are broad having intercellular spaces, larger towards central region. Sclereids are mostly solitary or in a group of 3 to 4 or more. Just below the cortex thick phloem having sieve tubes, companion cells and phloem fibers are present. The xylem consists of vessels, fibers, tracheids and xylem parenchyma. Vessels are mostly solitary or in groups of 2 or 3. The medullary rays are thick-walled, radiating, and parenchymatous. In longitudinal section vessels are seen vertically running broad segmented canals with bordered pitted and thickened wall, parenchyma are seen as narrow elongated tubes. The arrangements of medullary rays are horizontal, along with xylem vessels and fibers; tracheids are found along with the xylem vessels.

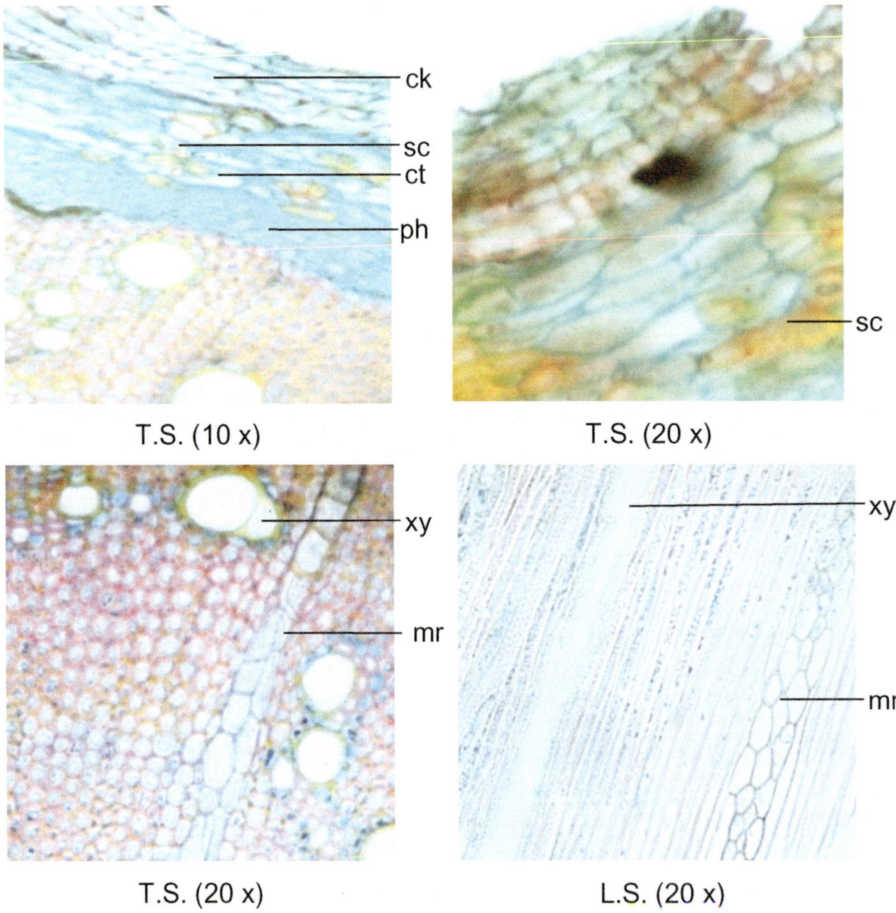

Transverse Section (T.S.) & Longitudinal Section (L.S.) of *Berberis umbellata* Root.

Abbreviation: ck- cork; ct- cortex; mr- medullary rays; ph- phloem; sc- sclereids; xy- xylem; xyv- xylem vessels.

HPTLC Studies

Working Solution

Accurately weighed (5g) crude powdered sample was subjected to cold extraction with methanol (50 ml), continuously shaken for 6h and then allowed to stand at room temperature (25 ± 2°C) for 18h. Extraction was repeated thrice, and the pooled filtrate was concentrate in rotatory evaporator (Buchi, Switzerland). Working solution for HPTLC study was made of strength 10 mg/ml.

Standard Solution

Berberine (≥ 98 per cent) was freshly prepared of strength 1 mg/ml.

Sample Application

2 µl of standard solution and 10 µl of sample was applied as 6 mm band with CAMAG HPTLC applicator on activated silica gel 60 GF_{254} HPTLC plate. The chromatogram was developed on pre-saturated CAMAG HPTLC twin trough chamber for 30 min and the solvent system was allowed to run for 8 cm.

Solvent System

n-propanol- water- formic acid (9:1:0.1 v/v).

Visualization

The chromatogram was visualized at UV_{366} and densitometric scanning of berberine was done at 350 nm.

Identification of Berberine

The standard berberine was identified at R_f 0.32.

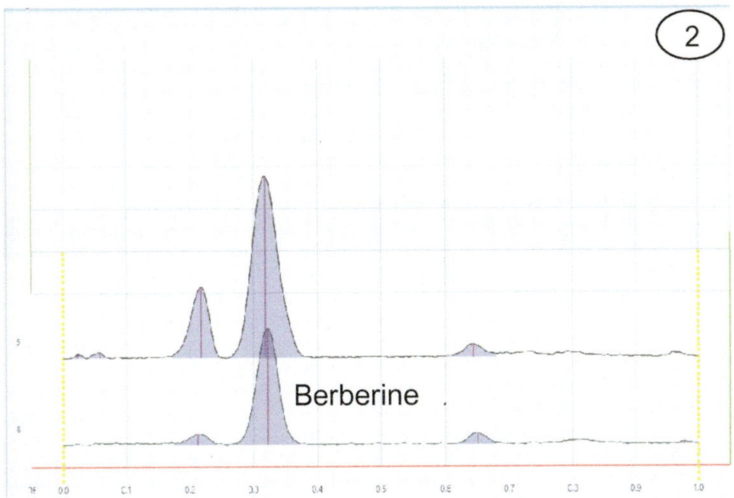

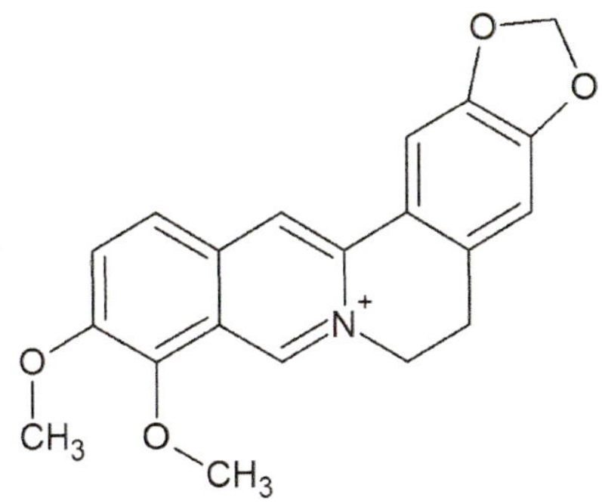

Berberine

Plate 1: HPTLC Chromatogram of *Berberis umbellata* (S) and Standard Berberine (R), Visualized at UV366 nm.
Plate 2: Overlay Spectra of Sample and Standard.

1. Singh R, Tiwari SS, Srivastava S, Rawat AKS (2012). Pharmacognostic evaluation of roots of *Berberis umbellata* Wall. ex G. Don. Indian J Nat Prod Res 3:55-60.

11

Bergenia ciliata (Haw.) Sternb

Family

Saxifragaceae

Vernacular Name

Sanskrit: Pashanbheda; Hindi: Pashanbhed, Dakachru; Bengali: Patharchuri; Marathi: Pashanbheda; English: Rock-foil; Kannada: Alepgaya, Pahanbhedi; Punjabi: Kachalu, Pashanbhed; Tamil: Sirupilai.

Distribution [1-3]

It is mostly distributed in the cold and temperate regions. It is found in the temperate Himalaya from Kashmir to Bhutan at an altitude of 900-3000 meters. In India it is reported from Lushai hills, West Bengal, Arunachal Pradesh, Meghalaya, Himalayas (Kumaon), Kyongnosla, Karponanag, Gangtok (Sikkim), and Almora (Uttarakhand).

Traditional Uses [4-8]

Juice of *B. ciliata* rhizome has been taken orally by human adult to relieve from intermittent fever (Malaria) and low fever (hypothermia).

Major Chemical Constituents [9-10]

Gallic acid, bergenin and catechin.

A Vegetative Plant

Dried Rhizome

Pharmacological Activity [11-14]

Antibacterial, antioxidant and anti-diabetic activity.

Plant Description

B. ciliata is a small perennial herb. In autumn season, it grows up to 35 cm, but it is well grown in moist and shady areas. Leaf margins are finely dentate with soft hairs; alternate, opposite and ex-stipulate. Flowers are pinkish white, solitary bisexual or rarely unisexual, calyx 5, corolla 4 or 5 and united. Fruits are capsular with many seeds. Rhizome is solid, cylindrical with small roots, root scar are often seen.

Anatomy

The transverse section of rhizome is almost circular in outline. The outermost zone is made of cork cell which are rectangular and 15 to 18 layered, followed by single layered cork cambium. Cortex multilayered, rectangular parenchymatous cells, some cells contain starch grain and rosette crystal of calcium oxalate, endodermis and pericycle absent. Vascular bundles are collateral, conjoint and open type; cambium is composed of 2-3 layers of thin walled cells. Xylem consist fibers, tracheids and vessels and phloem is composed of sieve element and parenchyma. Rosette crystals are found in whole section.

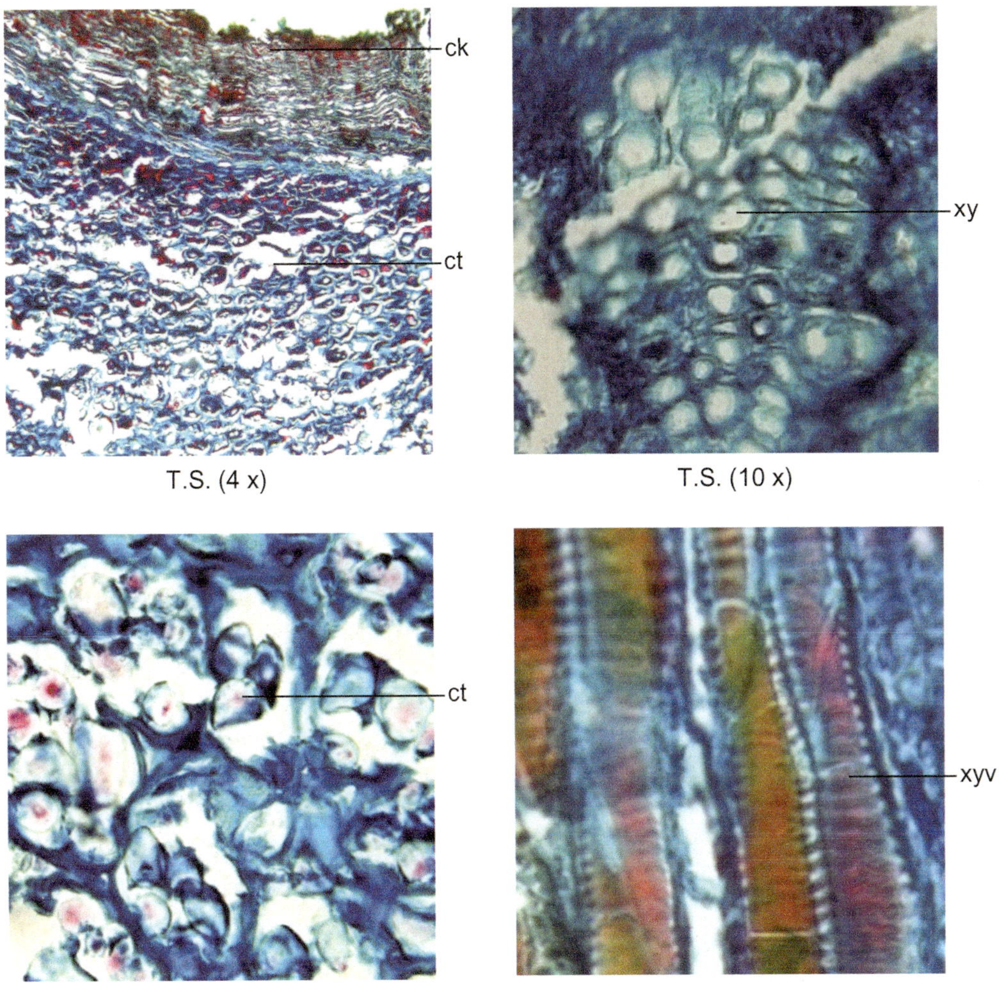

T.S. (20 x) L.S. (40 x)

Transverse Section (T.S.) & Longitudinal Section (L.S.) of *Bergenia ciliata* Rhizome.
Abbreviation: ck- cork; ct- cortex; cr- crystals; xy- xylem; xyv- xylem vessels.

HPTLC Studies

Working Solution

Accurately weighed (5g) crude powdered sample was subjected to cold extraction with methanol (50 ml), continuously shaken for 6h and then allowed to stand at room temperature (25 ± 2°C) for 18h. Extraction was repeated thrice, and the pooled filtrate was concentrate in rotatory evaporator (Buchi, Switzerland). Working solution for HPTLC study was made of strength 10 mg/ml.

Standard Solution

Gallic acid (>98 per cent) was prepared of strength 1 mg/ml.

Sample Application

2 µl of standard solution and 10 µl of sample was applied as 6 mm band with CAMAG HPTLC applicator on activated silica gel 60 GF_{254} HPTLC plate. The chromatogram was developed on pre-saturated CAMAG HPTLC twin trough chamber for 10 min and the solvent system was allowed to run for 8 cm.

Solvent System

Toluene: ethyl acetate: formic acid (6:6:1 v/v)

Visualization

The chromatogram was visualized at UV_{254} and densitometric scanning of gallic acid was done at 280 nm.

Identification of Gallic Acid[15]

The standard gallic acid was identified at R_f 0.5.

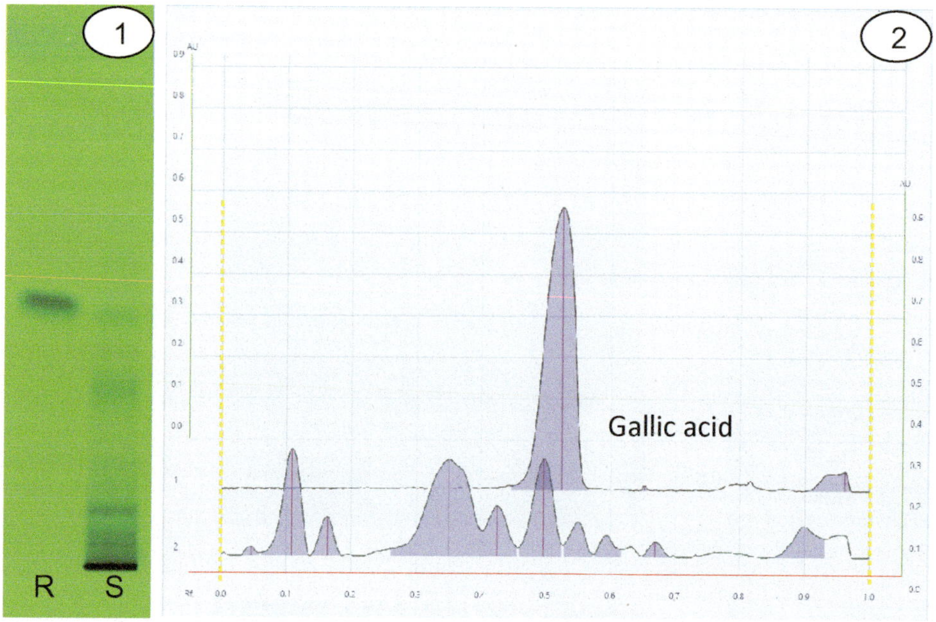

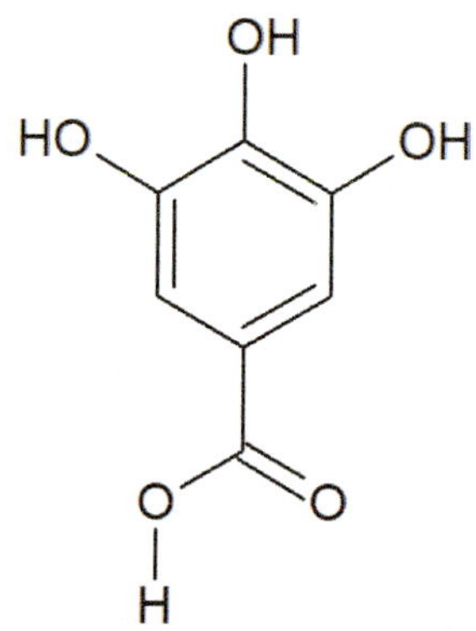

Gallic acid

Plate 1: HPTLC Chromatogram of *Bergenia ciliata* Sample (S) and Standard Gallic Acid (R).
Plate 2: Overlay Spectra of Sample and Standard.

1. Handa SS (1997). Indian Herbal Pharmacopoeia. Vol-1, Mumbai: A Joint Publication of RRL Jammu and IDMA 17-24.
2. Grierson AJ, Long DG (1983). Flora of Bhutan, including a record of plants from Sikkim, R. Bot. Gard. 1 (parts 1 and 2).
3. Hafidh R, Abdulamir A, Jahanshiri F, Abas F, Abu Bakar F, Sekawi Z (2009). Asia is the mine of natural antiviral products for public health. Open Complement Med J 1: 58-68.
4. Chauhan R, Ruby K, Dwivedi J (2012). *Bergenia ciliata* mine of medicinal properties: a review. Int J Pharm Sci Rev Res 15(2): 20–23.
5. Kumar A, Mitra M, Adhikari B, Rawat G (2015). Depleting indigenous knowledge of medicinal plants in cold-arid region of Nanda Devi Biosphere Reserve, Western Himalaya. Med Aromat Plants 4(195): 2167-0412.10001.
6. Hasan M, Gatto P, Jha P (2013). Traditional uses of wild medicinal plants and their management practices in Nepal—a study in Makawanpur district. Int J Med Aromat Plants 3: 102-112.
7. Bhattarai NK (1992). Folk herbal remedies of Sindhupalchok district, central Nepal. Fitoterapia 63(2):145 155.
8. Bhattarai NK (1992). Medicinal ethanobotany in the Karnali zone, Nepal. Economic botany 45(3):257-261.
9. Chauhan R, Ruby K, Dwivedi J (2013). Secondary metabolites found in *Bergenia* species: a compendious review, Reactions 15:17.
10. Kumar V, Tyagi D (2013). Review on phytochemical, ethnomedical and biological studies of medically useful genus *Bergenia*. Int J Curr Microbiol App Sci 2(5): 328-334.

11. Rajkumar V, Guha G, Kumar RA, Mathew L (2010). Evaluation of antioxidant activities of *Bergenia ciliata* rhizome. Records of Natural Products 4(1):38-48.
12. Sinha S, Murugesan T, Maiti K, Gayen J, Pal B, Pal M, Saha B (2001). Antibacterial activity of *Bergenia ciliata* rhizome. Fitoterapia 72(5): 550–552.
13. Ruby K, Chauhan R, Sharma S, Dwivedi J (2012). Polypharmacological activities of *Bergenia* species, Int J Pharmace Sci 1:100-109.
14. Bhandari MR, Jong-Anurakkun N, Hong G, Kawabata J (2008). α-Glucosidase and α-amylase inhibitory activities of Nepalese medicinal herb Pakhanbhed (*Bergenia ciliata*, Haw.) Food Chemistry 106(1): 247-252.
15. Srivastava S, Rawat A. (2007). Simultaneous determination of bergenin and gallic acid in different *Bergenia* species. JPC-Journal of Planar Chromatography-Modern TLC. Sep 1; 20(4): 275-7.

12

Bergenia ligulata Wall

Family

Saxifragaceae

Vernacular Name [1]

Hindi: Pashanbheda, Silparo; Bengali: Himasagara, Patranku; Gujarati: Pakhanbheda; Marathi: Pashanbheda; Kannada: Alepgaya, Pahanbhedi, Hittaga; Punjabi: Kachalu, Pashanbhed, Batpia; Tamil: Sirupilai; Urdu: Kachalu.

Distribution [2]

B. ligulata is distributed in the temperate Himalaya (from Kashmir to Nepal) from 2000-2700 m. It is also found in West Bengal, Uttarakhand and some states of North-East region.

Traditional Uses [3]

In Ayurveda and Unani medicines, the roots of B. ligulata posses laxative, analgesic, abortifacient properties and also used in treatment of vesicular calculi, urinary discharges, spleen enlargement and heart diseases.

A Vegetative Plant

Dried Pieces of Rhizome

Major Chemical Constituents [4-5]

Gallic acid, bergenin, (+)-catechin-7-O-β-D-glucopyranoside, 11-Ogalloyl bergenin and β-Sitosterol.

Pharmacological Activity [6-8]

Diuretic, hepatoprotective and anti-inflammatory activity.

Plant Description

B. ligulata is a small perennial herb with fleshy and short stem. It grows up to 50 cm, well grown in moist and shady areas. Leaf margins are finely dentate with soft hairs; alternate, opposite and ovate or round. Flowers are pink or purple, white, solitary bisexual or rarely unisexual, calyx 5, corolla 4 or 5 and united. Root is solid, cylindrical with small roots, root scar are often seen. The outer surface of rhizome is brown with aromatic odor and bitter taste.

Anatomy

The transverse section of rhizome is almost circular in outline. The outermost zone is made of cork cell which are rectangular, brown colored cell, followed by single layered cork cambium and 2-3 layers secondary cortex. Cortex multilayered, rectangular parenchymatous cells, some cells contain starch grain and rosette crystal of calcium oxalate, endodermis and pericycle absent. Vascular bundles are arranged in a ring, collateral, conjoint and open type; cambium is composed of 2-3 layers of thin walled cells. Xylem consist fibers, tracheids and vessels and phloem is composed of sieve element and parenchyma. Rosette crystals are found in whole section.

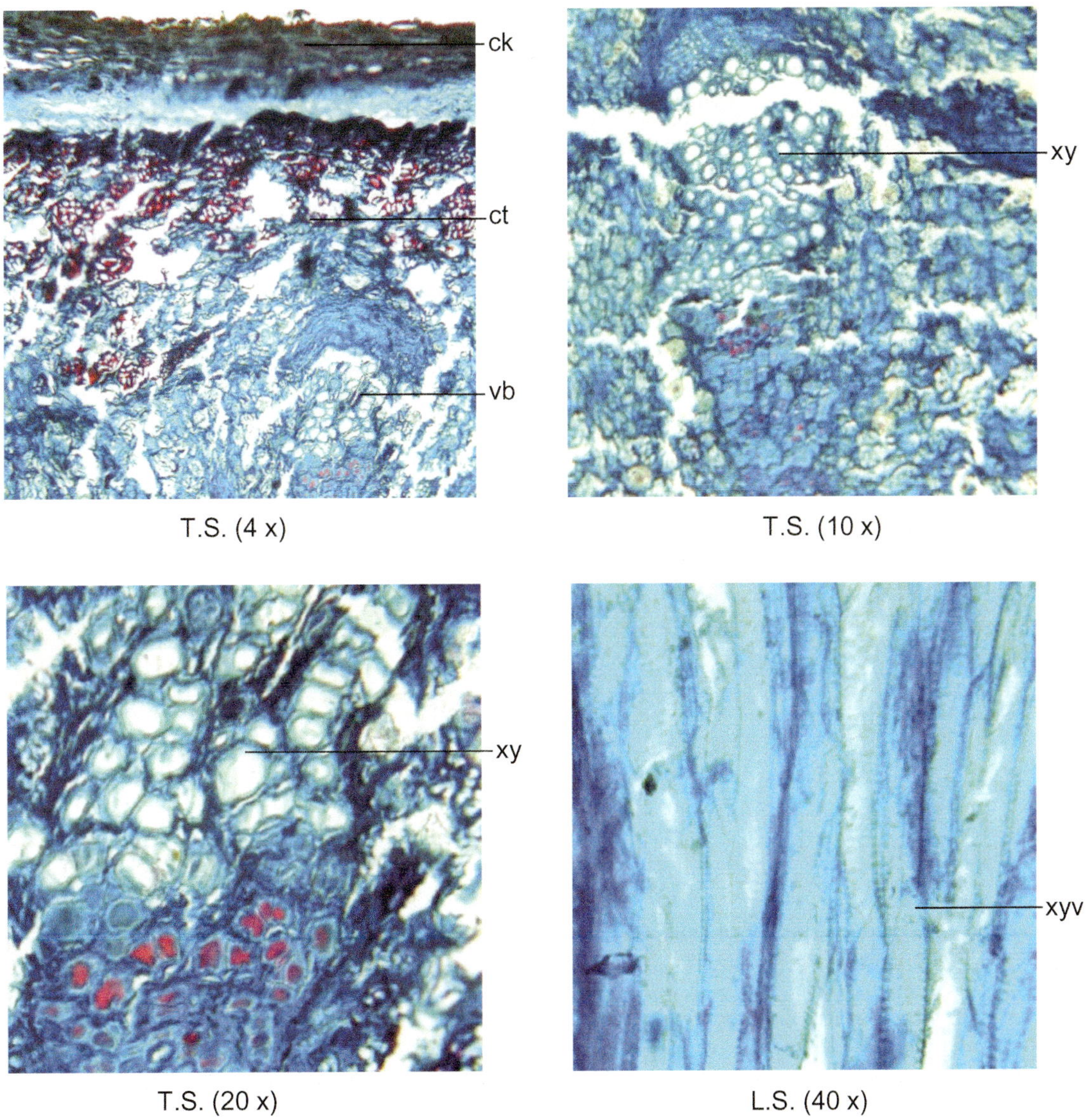

Transverse Section (T.S.) & Longitudinal Section (L.S.) of *Bergenia ligulata* Rhizome.

Abbreviation: ck- cork; ct- cortex; ph- phloem; vb- vascular bundle; xy- xylem; xyv- xylem vessels.

HPTLC Studies

Working Solution

Accurately weighed (5g) crude powdered sample was subjected to cold extraction with methanol (50 ml), continuously shaken for 6h and then allowed to stand at room temperature (25 ± 2°C) for 18h. Extraction was repeated thrice, and the pooled filtrate was concentrate in rotatory evaporator (Buchi, Switzerland). Working solution for HPTLC study was prepared of concentration 10 mg/ml.

Standard Solution

Gallic acid (>98 per cent) of strength 1mg/ ml was freshly prepared.

Sample Application

2 µl of standard solution and 10 µl of sample was applied as 6 mm band with CAMAG HPTLC applicator on activated silica gel 60 GF_{254} HPTLC plate. The chromatogram was developed on pre-saturated CAMAG HPTLC twin trough chamber for 10 min and the solvent system was allowed to run for 8 cm.

Solvent System

Toluene: ethyl acetate: formic acid (6:6:1 v/v)

Visualization

The chromatogram was visualized at UV_{254} and densitometric scanning of gallic acid was done at 280 nm.

Identification of Gallic Acid[9]

The standard gallic acid was identified at R_f 0.5.

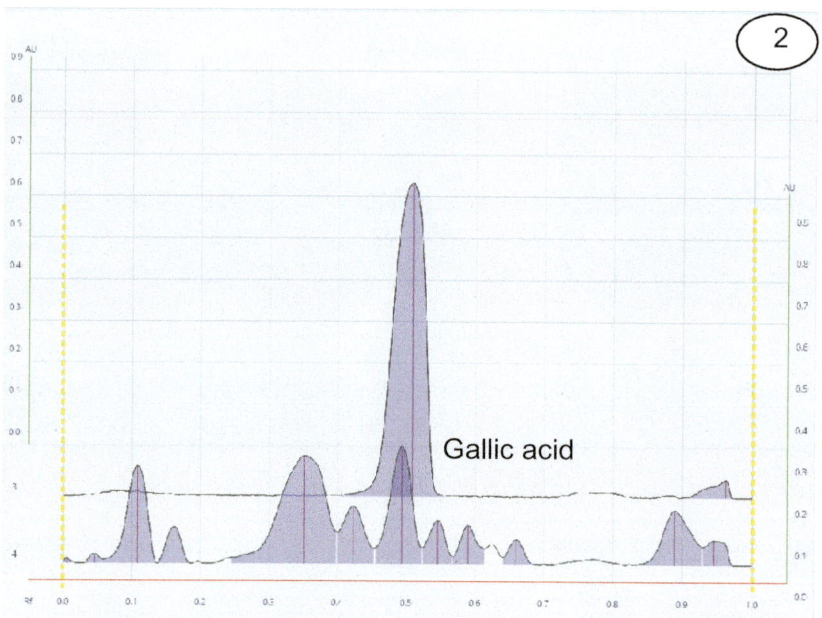

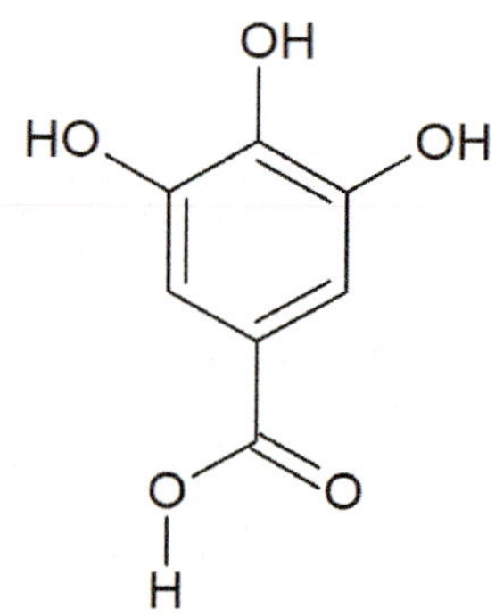

Gallic acid

Plate 1: HPTLC Chromatogram of *Bergenia ligulata* Sample (S) and Standard Gallic Acid (R).
Plate 2: Overlay Spectra of Sample and Standard.

1. Kirtikar K, Basu B (2005). Textbook of Indian Medicinal Plants. Volume II, 2nd ed. Dehradun, India: International Book Distributors 993-994.
2. Pandey G (1995). Medicinal Plants of Himalaya. Vol-I, Delhi, India: Sri Sadguru Publications 167-168.
3. Ruby KM, Dwivedi J, Chauhan R (2012). Pashanbheda a golden herb of Himalaya: a review. I j p r r 2(2): 97-105.
4. Umashankar D, Chawla A, Deepak M, Singh D, Handa S (1999). High pressure liquid chromatographic determination of bergenin and (+)- afzelechin from different parts of Paashaanbheda (*Bergenia ligulata*). Phytochemical Analysis 10: 44.
5. Jain MK, Gupta RJ (1962). Isolation of bergenin from *Saxifraga ligulata* Wall. Indian Chemical Society 39:559-560.
6. Umashankar DC (1997). Phytochemical and anti-inflammatory investigations of *Bergenia ligulata*. Ph.D. thesis, Punjab University, Chandigarh.
7. Shukla DS, RavishankarVJ, Bhavasar B (1996). Preliminary study on the hepatoprotective activity

of methanolic extract of *Paederia foetida* leaf, Fitoterapia, LX VII (2): 106-109.

8. Naik SR, Kalyanpur SN, Sheth UK (1972). Effect of antiinflammatory drugs on glutithione levels and liver succinic dehydrogenase activity in carrageenan edema and cotton pellet granuloma in rat. Biochemical Pharmacology 21: 511-5.

9. Srivastava S, Rawat A. (2007). Simultaneous determination of bergenin and gallic acid in different *Bergenia* species. JPC-Journal of Planar Chromatography-Modern TLC. Sep 1; 20(4): 275-7.

13

Bergenia stracheyi (Hook.f. and Thorns.)

Family

Saxifragaceae

Vernacular Name [1]

Hindi: Pashanbheda, Silparo; Bengali: Himasagara, Patranku; Gujarati: Pakhanbheda; Marathi: Pashanbheda; Kannada: Alepgaya, Pahanbhedi, Hittaga; Punjabi: Kachalu, Pashanbhed, Batpia; Tamil: Sirupilai; Urdu: Kachalu.

Distribution [2]

Bergenia stracheyi is found in the Western Himalayas, from 2700–4700 m, India, Afghanistan and Tajikistan. In India it is found in Uttarakhand, Himanchal Pradesh and Jammu and Kashmir.

Major Chemical Constituents [3]

Gallic acid, bergenin and Caryophyllene.

Pharmacological Activity [4–5]

Antioxidant and antifungal activity.

Plant Description

Bergenia stracheyi is a small perennial herb with

A Flowering Stalk

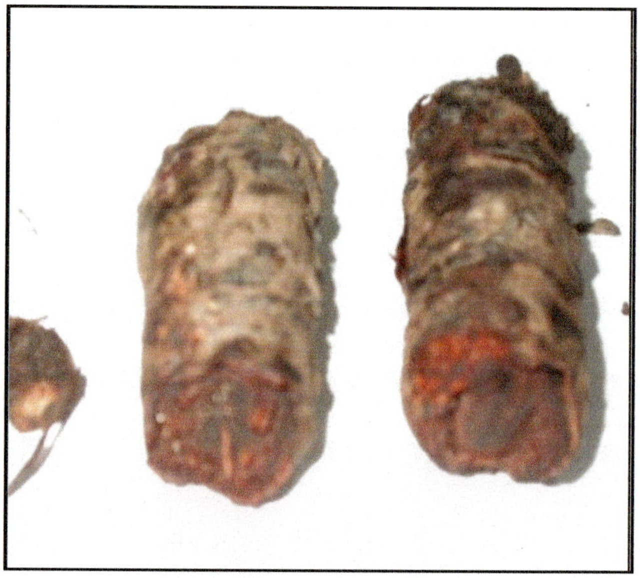

Dried Pieces of Rhizome

fleshy and short stem. It grows up to 50 cm, well grown in moist and shady areas. Leaf margins are finely dentate with soft hairs; alternate, opposite and ovate or round. Flowers are pink or purple, white, solitary bisexual or rarely unisexual, calyx 5, corolla 4 or 5 and united. Root is solid, cylindrical with small roots, root scar are often seen. The outer surface of rhizome is brown with aromatic odor and bitter taste.

Anatomy

The transverse section of rhizome is almost circular in outline. The outermost zone is made of cork cell which are rectangular, brown colored cell, 5-8 layers, followed by single layered cork cambium and 2-3 layers secondary cortex. Cortex multilayered, rectangular parenchymatous cells, some cells contain starch grain and rosette crystal of calcium oxalate, endodermis and pericycle absent. Vascular bundles are arranged in a ring, collateral, conjoint and open type; cambium is composed of 2-3 layers of thin walled cells. Xylem consist fibers, tracheids and vessels and phloem is composed of sieve element and parenchyma. Rosette crystals and starch grains are found in whole section.

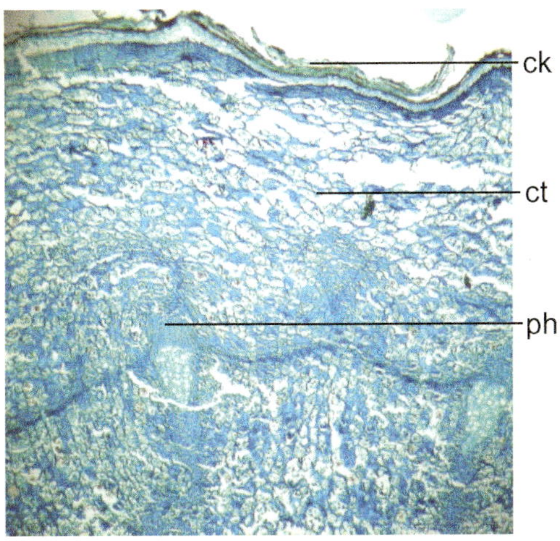

T.S. (4 x)

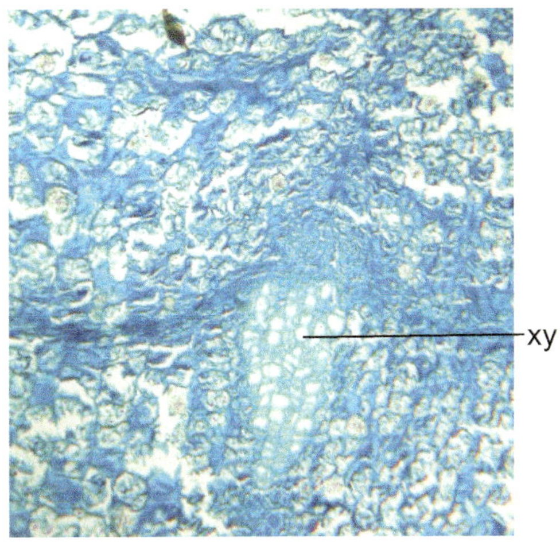

T.S. (20 x)

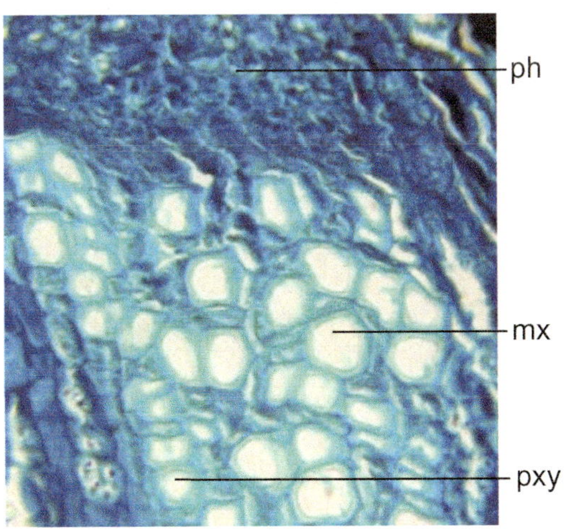

T.S. (20 x)

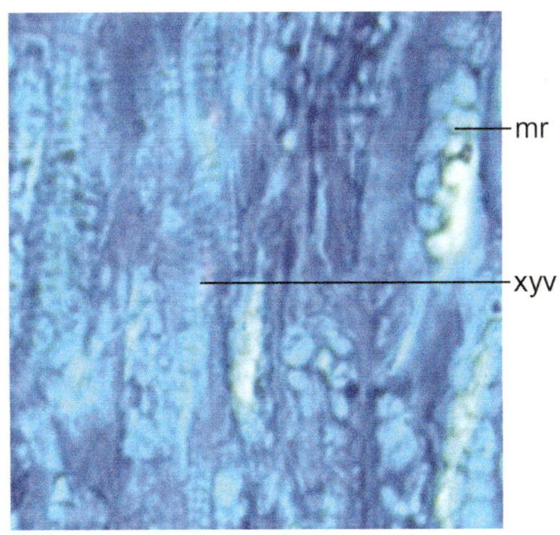

L.S. (40 x)

Transverse Section (T.S.) & Longitudinal Section (L.S.) of *Bergenia stracheyi* Rhizome.

Abbreviation: ck- cork; ct- cortex; mr- medullary rays; mx- meta xylem; ph- phloem; pxy- proto xylem; xy- xylem; xyv- xylem vessels.

HPTLC Studies

Working Solution

Accurately weighed (5g) crude powdered sample was subjected to cold extraction with methanol (50 ml), continuously shaken for 6h and then allowed to stand at room temperature ($25 \pm 2°C$) for 18h. Extraction was repeated thrice, and the pooled filtrate was concentrate in rotatory evaporator (Buchi, Switzerland). Working solution of 10 mg/ml strength was prepared for HPTLC study.

Standard Solution

Gallic acid (>98 per cent) was prepared in fresh 1 mg/ml concentration

Sample Application

2 µl of standard solution and 10 µl of sample was applied as 6 mm band with CAMAG HPTLC applicator on activated silica gel 60 GF_{254} HPTLC plate. The chromatogram was developed on pre-saturated CAMAG HPTLC twin trough chamber for 10 min and the solvent system was allowed to run for 8 cm.

Solvent System

Toluene: ethyl acetate: formic acid (6:6:1 v/v)

Visualization

The chromatogram was visualized at UV_{254} and densitometric scanning of gallic acid was done at 280 nm.

Identification of Gallic Acid[6]

The standard gallic acid was identified at R_f 0.5.

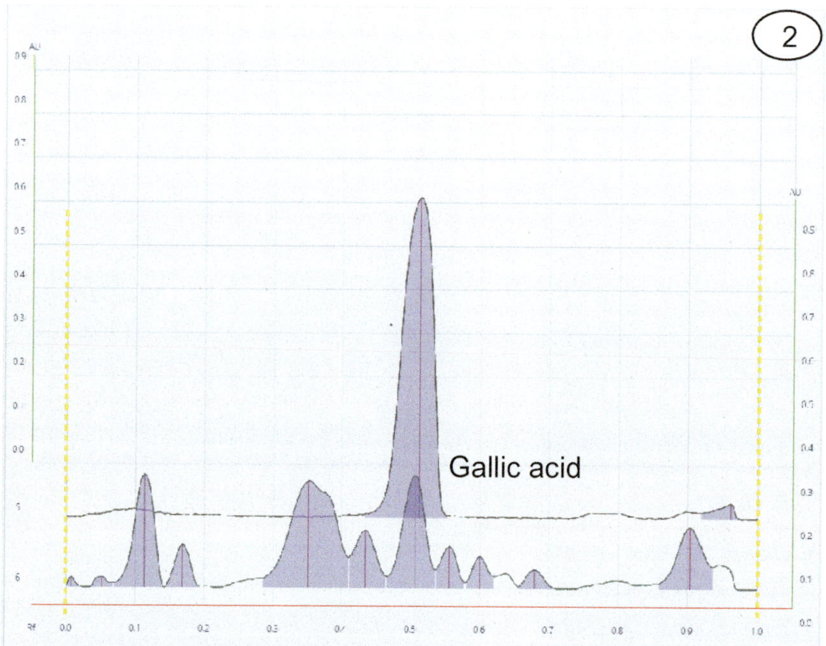

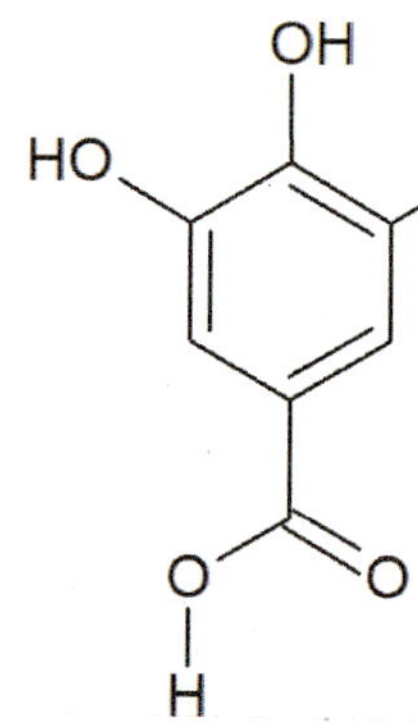

Gallic acid

Plate 1: HPTLC Chromatogram of *Bergenia stracheyi* Sample (S) and Standard Gallic Acid (R).
Plate 2: Overlay Spectra of Sample and Standard.

1. *Bergenia stracheyi* on www.efloras.org
2. http://suite101.com/article/bergenin-a103299
3. Zhao JY, Liu JM, Zhang XY, Liu Z.J, Tsering T, Zhong Y, Nan P (2006). Chemical composition of the volatiles of three wild *Bergenia* species from western China. Flavour Fragrance J 21(3): 431-434.
4. Kumar V, Tyagi D (2013). Phytochemical screening and free-radical scavenging activity of *Bergenia stracheyi*. J Pharmacog Phyto 2(2).
5. Kumar V, Tyagi D (2013). Antifungal activity evaluation of different extracts of *Bergenia stracheyi*. Int J Curr Microbiol App Sci 2(7): 69-78.
6. Srivastava S, Rawat A. (2007). Simultaneous determination of bergenin and gallic acid in different *Bergenia* species. JPC-Journal of Planar Chromatography-Modern TLC. Sep 1; 20(4): 275-7.

14

Boerhavia diffusa Linn.

Family

Nyctaginaceae

Vernacular name [1]

Sanskrit: Punarnava, Raktakanda; Hindi: Snathikari; Bengali: Punarnnava; Marathi: Ghetuli; English: Spreading Hogweed, Red Hogweed; Punjabi: Khattan; Tamil: Chattarani; Assamese: Ranga Punarnabha; Kannada: Kommegida.

Distribution [2-3]

The genus *Boerhavia* consist of 40 species, distributed in tropical and subtropical regions of the World. Six species *viz. B. diffusa, B. erecta, B. rependa, B. chinensis, B. hirsute* and *B. rubicunda* are found in India. The most commonly available species is *B. diffusa* distributed throughout the warmer parts of the country up to 2,000 m.

Traditional Uses [3-6]

The plant is diuretic, aphrodisiac, cardiac stimulant, diaphoretic, expectorant, febrifuge and laxative. Root decoction is used to treat fever and possesses anti-convulsant, expectorant, laxative and stomachic

A Vegetative Plant

Dried Root

properties. The root powder when mixed with mamira (*Thalictrum foliolosum*) is used to treat eye diseases.

Major Chemical Constituents [7-10]

Oleanolic acid, punarnavine, quercetin, kaempferol

Pharmacological Activity [11-13]

Anti-diabetic, anti-stress, anti-oxidant and hepato-protective activity

Plant Description

Boerhavia diffusa is a perennial crawling weed, spreading prostrate, up to 1 m long or more, having spreading branches. Stem is cylindrical, somehow glaberous, prostrate, swollen at nodes from which leaves arises. Leaves are opposite and unequal, glaberous, ovate-oblong, apex slightly pointed, rounded base, margin entire with long petiole. Flowers are very small, pink, stalk short or sessile, umbel arranged on long cylindrical stalk and terminal. Fruits are one seeded nut and rounded. Roots are yellowish brown to brown colored, soft, stout, fusiform with a woody root stock, root scars are often seen, no odor and slight bitter taste.

Anatomy

The transverse section of mature root is almost circular in outline. The outermost zone is made of cork cell which made up of thin walled tangentially elongated cells, followed by 1-2 layers of thin walled cork cambium. 2-3 layers of parenchymatous secondary cortex. Starch grains are rounded and oval shaped abundantly found in cortex and parenchymatous tissue between xylem elements. Cortex multilayered, oval to polygonal parenchymatous cells, several concentric xylem tissues are found just below the cortical region. Phloem patches are present outside of xylem vessels, made up of sieve element and parenchyma, rectangular shape arranged in radial rows. Xylem is composed of vessels, tracheids and fibers, vessels are mostly found in groups of 2-6. Center region of root is mostly occupied by primary vascular bundle. Raphides of calcium oxalate present in cortical and parenchymatous region between the xylem tissues.

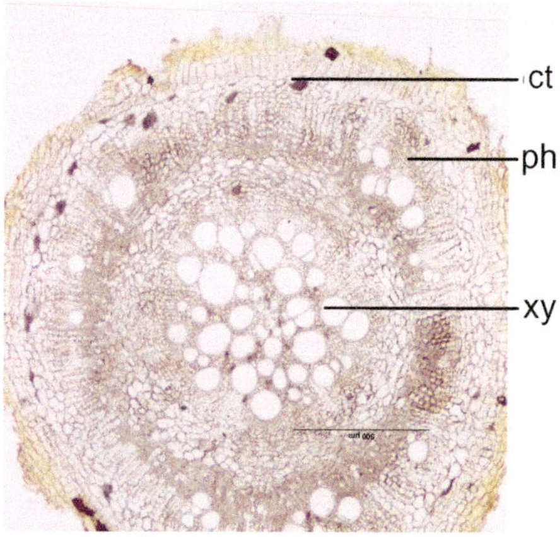

T.S. (4 x)

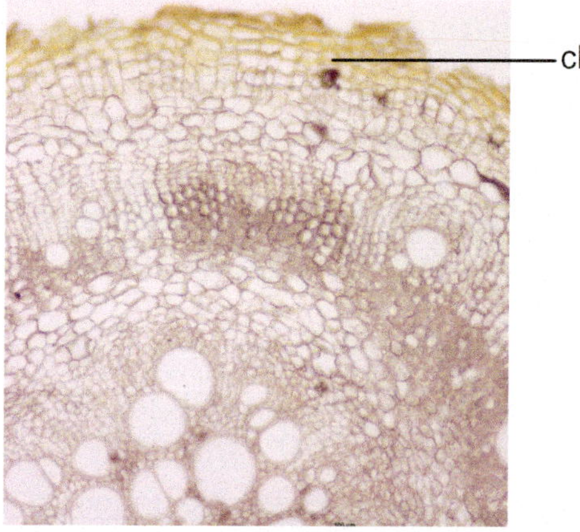

T.S. (10 x)

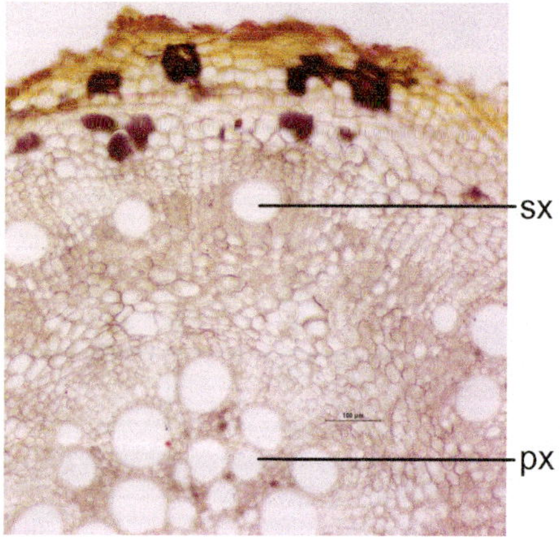

T.S. (10 x)

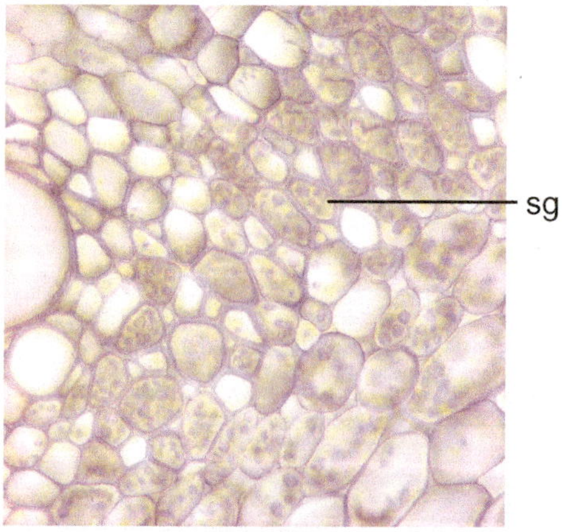

T.S. (40 x)

Transverse Section (T.S.) of *Boerhavia diffusa* Root.

Abbreviation: ck- cork; ct- cortex; ph- phloem; px- primary xylem; sx- secondry xylem; xy- xylem; xyv- xylem vessels; sg- starch grains.

HPTLC Studies

Sample Solution

Accurately weight (5 g) crude powdered sample was subjected to cold extraction with methanol (50 ml) and continuously shaken was done for 6 h then kept overnight at room temperature (25 °C ± 2). The extraction process was repeated thrice, filtered and the pooled filtrate was concentrate in rotatory evaporator (Buchi, Switzerland). Working solution of strength 10 mg/ml was freshly prepared on the day of quantative analysis.

Standard Solution

Oleanolic acid (≥ 98 per cent) was freshly prepared of strength 1 mg/ml.

Sample Application

2 µl of standard solution and 10 µl of sample was applied by CAMAG HPTLC applicator on HPTLC preset silica gel 60 GF_{254} plate. The 20 minute saturation was given and the plate was developed in saturated chamber up to 8 centimeters.

Solvent System

Toluene- ethyl acetate- methanol- formic acid (7:2:0.5:0.5 v/v).

Visualization

The chromatogram was visualized after derivatization with anisaldehyde sulphuric acid reagent and densitometric scanning of oleanolic acid was done at 530 nm.

Identification of Oleanolic Acid

The standard oleanolic acid was identified at R_f 0.35.

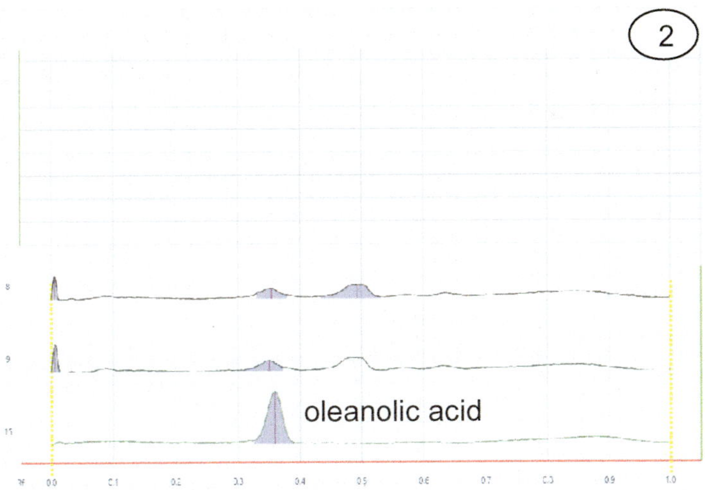

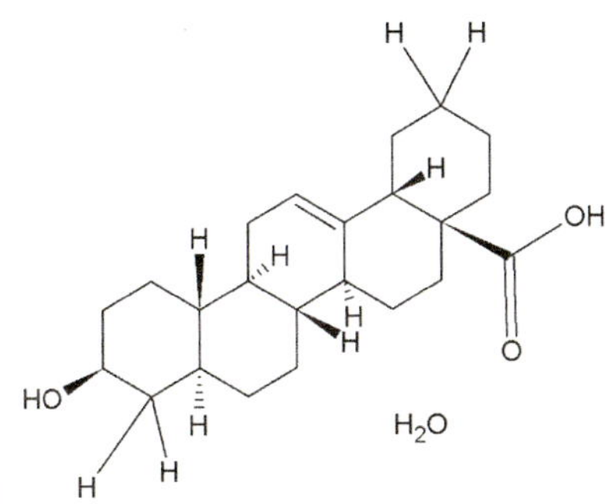

oleanolic acid

Plate 1: HPTLC Chromatogram of *Boerhavia diffusa* Sample (S) and Standard Oleanolic Acid (R).
Plate 2: Overlay Spectra of Sample and Standard.

1. Oudhia P. Traditional medicinal knowledge about useful herb Punarnava (*Boerhavia diffusa*, family Nyctaginaceae) in Chhattisgarh Plains, India, Available at: http://botanical.com/site/column_poudhia/138_punarnava.
2. Najam A, Singh AK, Verma HN (2008). Ancient and modern medicinal potential of *Boerhavia diffusa* and *Clerodendrum aculeatum*. Res Environ Life Sciences 1(1): 1-4.
3. Nayak P, Thirunavoukkarasu M (2016). A review of the plant *Boerhavia diffusa*: its chemistry, pharmacology and therapeutical potential. jour Phytopharma 5(2): 83-92.
4. Prajapati ND, Purohit SS, Sharma AK, Kumar T (2004). A handbook of Medicinal Plants-A complete source book, India.
5. Bedi SJ (1978). Ethnobotany of the Ratan Mahal hills, Gujarat, India. Econ Bot 32(3): 278-284.
6. Adesina SK (1979). Anticonvulsant properties of the roots of *Boerhavia diffusa*. Pharmaceutical Biol 17: 84-86.
7. Gupta RBL, Singh S, Dayal Y (1962). Effect of punarnava on the visual acuity and refractive errors. Ind J Med Res 50: 428-434.
8. Mishra AN, Tiwari HP (1971). Constituents of the roots of *Boerhavia diffusa*. Phytochemistry 10: 3318.
9. Agarwal RR, Dutt SS (1936). Chemical examination of punarnava or *Boerhavia diffusa* Linn. II. Isolation of an alkaloid punarnavine. Chemical Abstract 30: 3585.

10. Ferreres F, Sousa C, Justin M, Valentão P, Andrade PB, Llorach R, Rodrigues A, Seabra RM, Leitão A (2005). Characterization of the Phenolic Profile of *Boerhavia diffusa* L. by HPLC-PAD-MS/MS as a tool for Quality Control. Phytochem Anal 16: 451-458.

11. Bhatia V, Kinja K, Bishnoi H (2001). Antidiabetic Activity of the Alcoholic Extract of the Arial Part of *Boerhavia diffusa* in Rats. In and Gnaneshwari D editors. Rec Res Sci Tech 3(7): 04-07.

12. Rachh PR, Rachh MR, Modi DC (2009). In-vitro evaluation of antioxidant activity of punarnava (*Boerhavia diffusa* L.). Int Jour Pharma Res 1: 36-40.

13. Rawat AKS, Mehrotra S, Tripathi SC, Shome U (1997). Hepatoprotective activity of *Boerhavia diffusa* L. roots, a popular Indian ethnomedicine. Jour Ethnopharma 56: 61-66.

15

Calotropis gigantea (L.) R. Br.

Family

Asclepiadaceae

Vernacular name [1]

Sanskrit: Arka, manra; Hindi: Safed aak, Aak, Madar, Sveta Arka; English: Crown flower, giant Indian milkweed; Gujarati: Aakando; Bengali: Akanda; Tamil: Erukku (Arkkam); Kannada: Arkagida.

Distribution [2]

Calotropis gigantea (L.) R. Br. is drought resistant weed, commonly found throughout India. The species is also salt tolerant to a relatively high degree and is widely available in China, Malaysia, Africa and America.

Traditional Uses [3]

The dried root powder is used to cure bronchitis, asthma, leprosy, eczema and elephantiasis.

Major Chemical Constituents [4-6]

Lupeol, calotropin, α-amyrin, β-amyrin, β-sitosterol

Pharmacological Activity [7-11]

Anti-microbial, wound healing, Anti-pyretic, Insecticidal and anti-cancer activity.

A Flowering Stalk

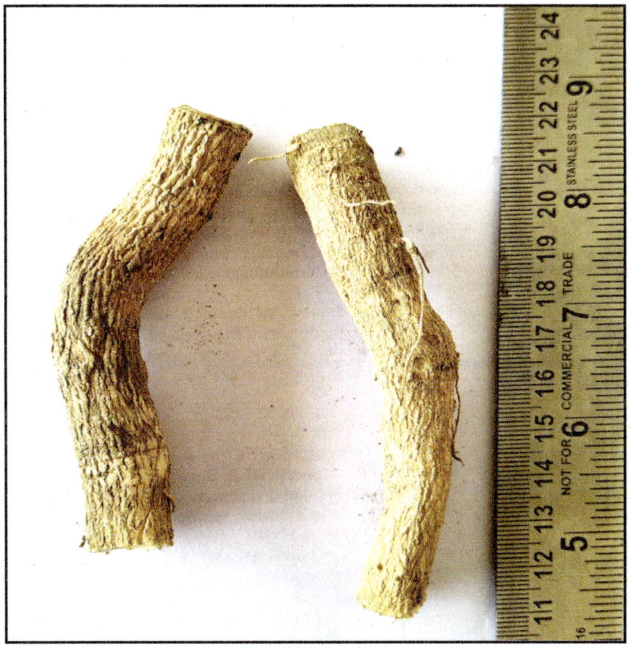

Dried Root

Plant Description

Calotropis gigantea is soft-wooded shrub, and occasionally a tree attained 6 m height. On cutting, all the parts of the plant extinguish the white milky latex. Leaves are large, opposite-decussate, sub sessile, extipulate. Stem is green cylindrical, hollow in centre. Flower is Bracteate, complete, bisexual, actinomorphic, pentamerous, and purple and has a faint odour. Sepals are 5, Polysepalous, Petals five, gamopetalous, Stamens five, gynandrous, anther dithecous and ovary Bicarpellary. Root is simple, thick, soft, branched, woody, smooth tap roots bearing a few secondary and tertiary roots.

Anatomy

Transverse section of root is circular in outline, show very narrow range of cork and central wood, encircled by narrow phloem and wider cortex. Detailed structure accounted that the root shows narrow cork made of 8 to 10 row of dark brown cells, followed by wide parenchymatous cortex, pericycle, narrow parenchymatous phloem and distinct cambium. Starch grains are rounded or oval in shape, are abundantly seen in the cortical cells, phloem parenchyma, medullary rays, and xylem fibers. Lignified medullary rays in continuation with that of secondary phloem are radially arranged with xylem vessels, xylem fibers. Secondary phloem occurs in broad continuous cylinder around the xylem. Xylem is surrounded with xylem fiber and conjunctive tissues. Secondary xylem is in the form of dense solid cylinder with even periphery. Xylem elements include thin walled. In longitudinal section vessels are seen vertically running broad segmented canals, parenchyma are seen as narrow elongated tubes with pitted walls, xylem fibers are vertically arranged. The arrangements of medullary rays are horizontal, along with xylem vessels and fibers.

HPTLC Studies

Sample Solution

Accurately weight (5 g) crude powdered sample was subjected to cold extraction with methanol (50

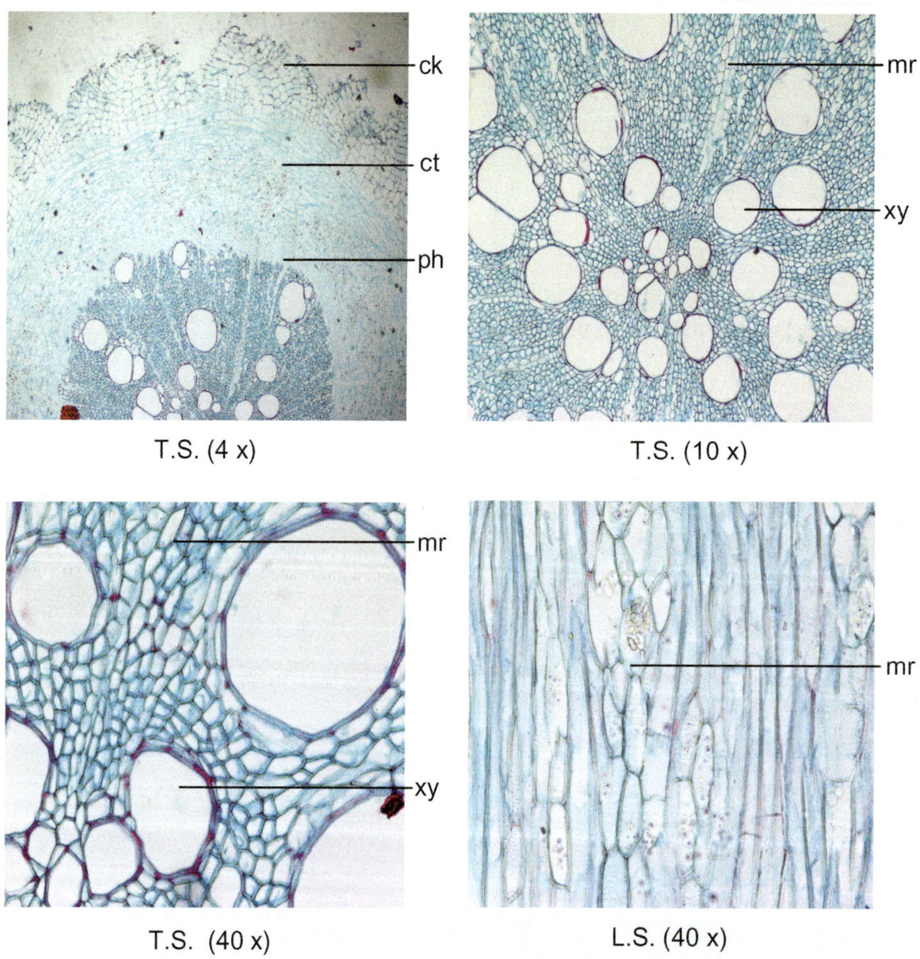

Transverse Section (T.S.) & Longitudinal Section (L.S.) of *Calotropis gigantea* Root.

Abbreviation: ck- cork; ct- cortex; mr- medullary rays; ph- phloem; xy- xylem; xyf-xylem fibre; xyv- xylem vessels.

ml), solution was continuously shaken for 6 h and then allowed to stand at room temperature (25 °C ± 2) for 18 hrs. The extraction process was repeated thrice, filtered and the pooled filtrate was concentrate in rotatory evaporator (Buchi, Switzerland). Working solution of 10 mg/ml was freshly prepared on the day of quantative analysis.

Standard Solution

Lupeol (≥ 98 per cent) was freshly prepared of strength 1 mg/ml.

Sample Application

2 µl of standard solution and 10 µl of sample was applied by CAMAG HPTLC applicator on HPTLC preset silica gel 60 GF_{254} plate. The 20 minute saturation was given and the plate was developed in saturated chamber up to 8 centimeters.

Solvent System

Toluene- ethyl acetate- methanol- formic acid (7:2:0.5:0.5 v/v).

Visualization

The chromatogram was visualized after derivatization with anisaldehyde sulphuric acid reagent and densitometric scanning was done at 530 nm

Identification of Lupeol

The standard lupeol was identified at R_f 0.62.

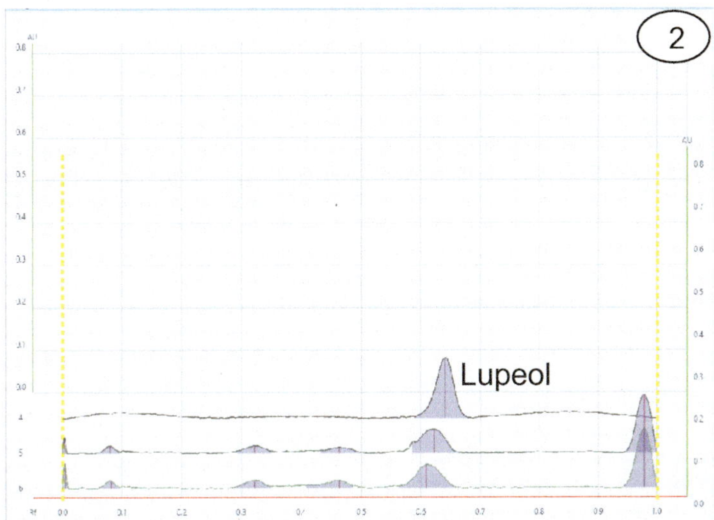

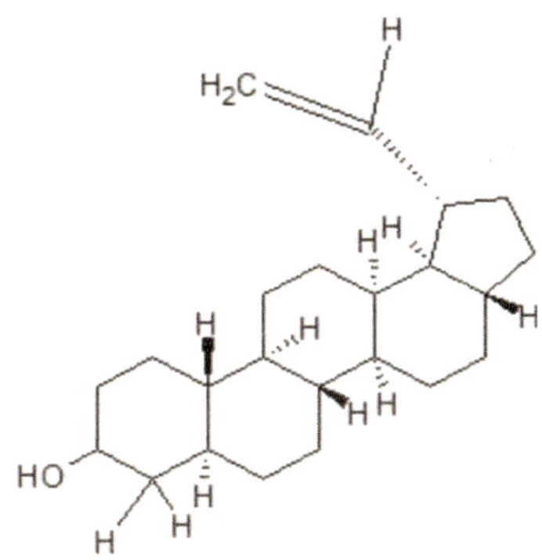

Lupeol

Plate 1: HPTLC Chromatogram of *Calatropis gigantea* Sample (S) and Standard Lupeol (R) Visualized after Derivatization with Anisaldehyde Sulphuric Acid reagent.

Plate 2: Overlay Spectra of Sample and Standard.

1. In: Compendium of Indian Medicinal Plants. Central Drug Research Institute, Lucknow and Publications and Information Directorate, N. Delhi. 70-73.
2. Gamble J S (1935). Flora of the Presidency of Madras Vol. I, II, III, Botanical survey of India, Calcutta.
3. Warrier, P.K., V.P.K Nambiar, and C. Mankutty 1994. Indian Medicinal Plants. Orient Longman; Chennai, India p. 341-345.
4. Habib RM, Nikkon F, Rahaman M (2007). Isolation of β-Sitosterol from methanolic extract of root bark of *Calotropis gigantea* (Linn). Pak J Biol Sci 1-3.
5. Kupchan SM, Knox JR, Kelsey JE, Renauld JAS (1964). Calotropin, a cytotoxic principle isolated from *Asclepias curassavica* L. Science 146: 1685-1686.
6. Sen S, Sahu NP and Mahato SB (1992). Flavonol glycosides from *Calotropis gigantean*. Phytochemistry 31: 2919-2921.
7. Alam MA, Habib MR, Nikkon R, Rahman M, Karim MR (2008). Antimicrobial activity of akanda (*Calotropis gigantea* L.) on some pathogenic bacteria. Bangladesh J Sci Ind Res 43(3): 397-404.
8. Alam MA, Habib MR, Nikkon F, Khalequzzaman M, Karim MR (2009). Insecticidal activity of root bark of *Calotropis gigantea* L. against *Tribolium castaneum* (Herbst). World J Zoology 4(2): 90-95.

9. Chitme HR, Chandra R, Kaushik S (2005). Evaluation of antipyretic activity of *Calotropis gigantea* (Asclepiadaceae) in experimental animals. Phototherapy Res 19(5): 454-456.

10. Deshmukh PT, Fernandes J, Aarte A, Toppo E (2009). Wound healing activity of *Calotropis gigantea* root bark in rats. J Ethnopharmacol 125(1): 178-181.

11. Vishnu P, Jain P, Meenashi Vanathi B, Vasanth Raj P, Venkatesh B, Kamath J (2015). Methanolic Root Extract of *Calotropis gigantea* Induces Apoptosis in Human Hepatocellular Carcinoma by Altering Bax/Bcl-2 Expression. Ameri J Phar Sci 3(1): 13-17.

16

Calotropis procera (Ait.) R.Br.

Family

Asclepiadaceae.

Vernacular Name [1]

Sanskrit: Arka, Alaka; Hindi: Aaka, Ak, Madar; English: Crown flower, Sodom Apple, Roasted tree.

Distribution [2]

Calotropis procera (Ait.) R.Br. is a common weed found throughout India. It is a native of India, Pakistan, China, Nepal, Iran, Israel, Zimbabwe and Nigeria.

Traditional Uses [3–5]

The root powder when mixed with goat milk is used in epilepsy. The root bark is used to treat elephantiasis and chronic cases of dyspepsia, constipation, indigestion and mucus in stools.

Major Chemical Constituents [6–7]

Lupeol, calotropin, α-amyrin, β-amyrin, β-sitosterol.

Pharmacological Activity [7–8]

Antifertility, wound healing and anticonvulsant property.

A Flowering Stalk

Dried Piece of Root

Plant Description

Calotropis procera is soft-wooded shrub, occasionally tree and can attained 4 to 5 m height. The root is simple, branched, woody, covered with corky bark and secondary roots are arises in deep soil. The leaves are slightly leathery, large, opposite-decussate, sub sessile, ovate-oblong and pubescent. Stem is green cylindrical, hollow in centre, covered with fibrous bark which is sometime used as rope. Flowers are shallow, bell-shaped, complete, bisexual, actinomorphic, pentamerous, hypogynous, purple in color and have a faint odour. Sepals are 5, poly-sepalous, five petals, lobed, gamopetalous, stamens are five, gynandrous, dithecous anther and bicarpellary ovary. Fruit is simple, fleshy sub-globose to obliquely ovoid follicle.

Anatomy

Transverse section of root has distinct and continuous periderm, wide cortex and circular vascular cylinder. Detailed structure accounted that the root shows narrow cork made of 3 to 5 row of dark brown cells, followed by wide parenchymatous cortex, pericycle phloem narrow parenchymatous and distinct cambium. Starch grains are rounded or oval in shape, are abundantly seen in the lower cortical cells or cortical cells which present near to phloem. Upper 10 -12 layers of cortical cells have no or very few starch grains. Lignified medullary rays in continuation with that of secondary phloem are radially arranged with xylem vessels, xylem fibres and conjunctive tissues. Secondary phloem occurs in broad continuous cylinder around the xylem. Xylem is surrounded with xylem fibres and conjunctive tissues. Secondary xylem is in the form of dense solid cylinder with even periphery. In longitudinal section vessels are seen vertically running broad segmented canals, parenchyma are seen as narrow elongated tubes with pitted walls, xylem fibres are vertically arranged. The arrangements of medullary rays are horizontal, along with xylem vessels and fibres.

HPTLC Studies

Sample Solution

Accurately weight (5 g) crude powdered sample was subjected to cold extraction with methanol (50 ml) and

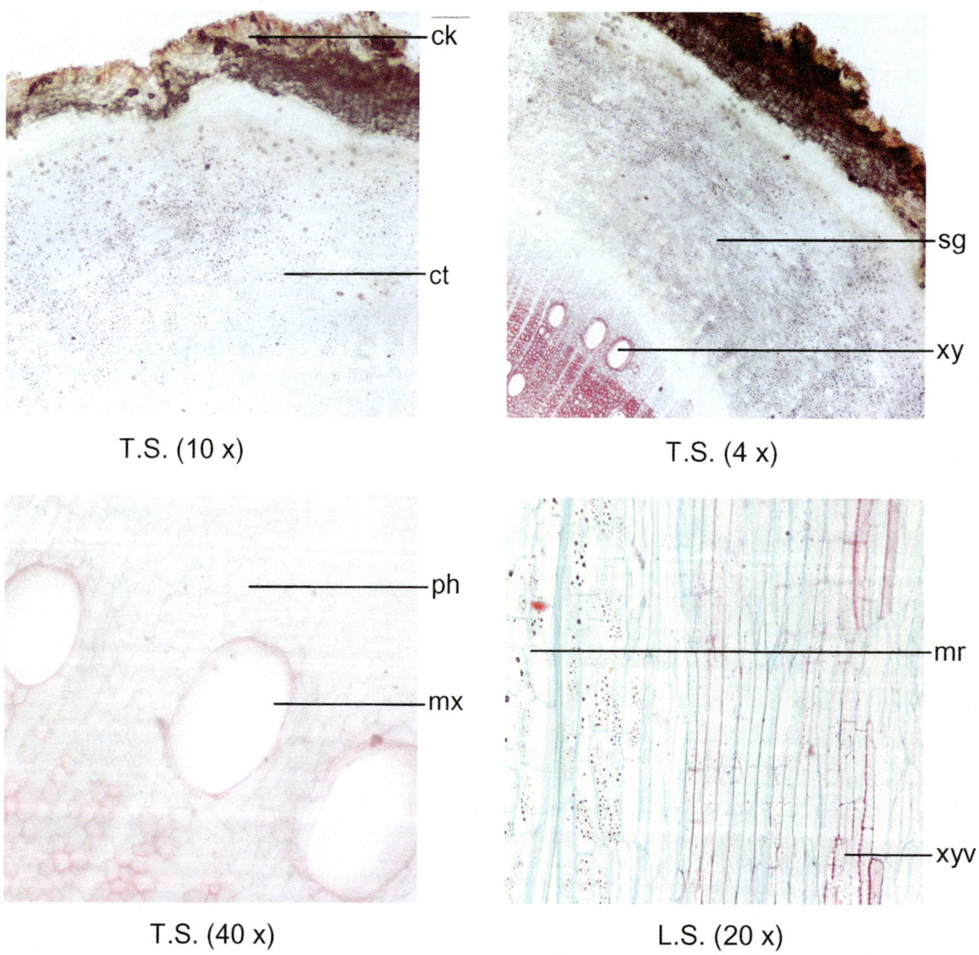

Transverse Section (T.S.) & Longitudinal Section (L.S.) of *Calotropis procera* Root.

Abbreviation: ck- cork; ct- cortex; mr- medullary rays; mx- metaxylem; ph- phloem; sg- starch grains; xy- xylem; xyv- xylem vessels.

continuously shaken was done for 6 h then allowed to stand at room temperature (25 °C ± 2) for 18 hrs. The extraction process was repeated thrice, filtered and the pooled filtrate was concentrate in rotatory evaporator (Buchi, Switzerland). Working solution of strength 10 mg/ml was freshly prepared on the day of quantative analysis.

Standard Solution

Lupeol (≥ 98 per cent) was freshly prepared of strength 1 mg/ml.

Sample Application

2 µl of standard solution and 10 µl of sample was applied by CAMAG HPTLC applicator on HPTLC preset silica gel 60 GF_{254} plate. The 20 minute saturation was given and the plate was developed in saturated chamber up to 8 centimeters.

Solvent System

Toluene- ethyl acetate- methanol- formic acid (7:2:0.5:0.5 v/v).

Visualization

The chromatogram was visualized after derivatization with anisaldehyde sulphuric acid reagent and densitometric scanning of lupeol was done at 530 nm.

Identification of Lupeol[9]

The standard lupeol was identified at R_f 0.62.

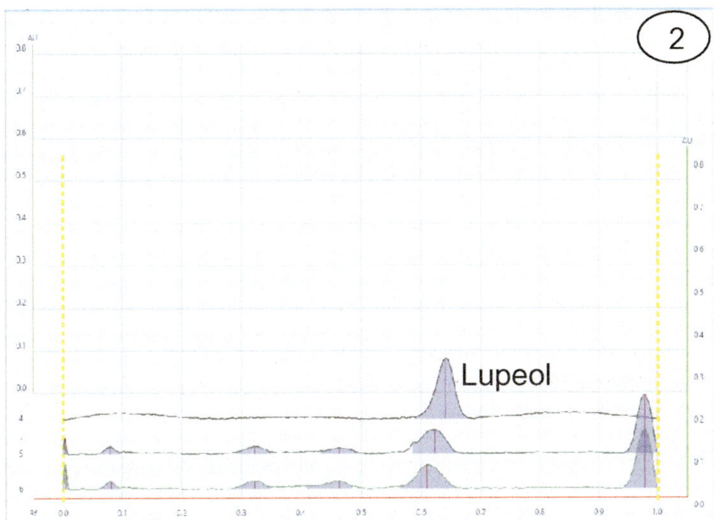

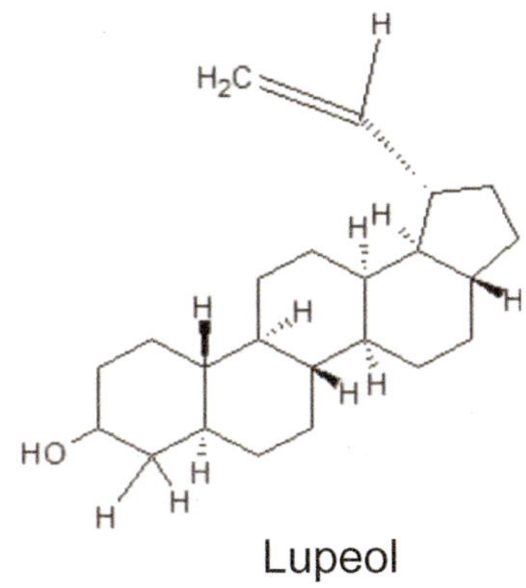

**Plate 1: HPTLC Chromatogram of *Calatropis gigantia* Sample (S) and Standard Lupeol (R) .
Plate (2) Overlay Spectra of Sample and Standard.**

1. Ahmed KK, Rana AC, Dixit VK (2005). *Calotropis* species (Ascelpediaceae): A comprehensive review. Pharma Mag 1: 48-52
2. Chatterjee A, Chandra PS (1995). The treatise of Indian medicinal plants. Vol. 4, CSIR, New Delhi; 130.
3. Verma R, Satsangi GP, Shrivastava JN (2010). Ethno-medicinal profile of different plant parts of *Calotropis procera* (Ait.) R.Br. Ethnobot Leaf 14: 721-42.
4. Sharma K, Kharb R, Kaur R (2011). Pharmacognostical aspects of *Calotropis procera* (Ait.) R.Br. Int J Pharm Bio Sci 2: 1-9.
5. Kumar VL, Arya S (2006). Medicinal uses and pharmacological properties of *Calotropis procera*. Rec Pro Med Pla 373-388.
6. Atef GH, Elgamal MHA, Morsy NAM, Duddeck H, Kovacs J, Toth G (1999). Two cardenolides from *Calotropis procera*. J Magn Reson Chem 17: 754-757.
7. Quazi S, Mathur K, Arora S (2013). *Calotropis procera*: An overview of its phytochemistry and pharmacology. Indian J Drugs 1: 63-9.
8. Tsala DE, Nga N, Thiery BN, Bienvenue MT, Theophile D (2015). Evaluation of the antioxidant activity and the healing action of the ethanol extract of *Calotropis procera* bark against surgical wounds. J Inter Ethnopharma; 4:64-9.
9. Sharad Srivastava, Adarsh Pratap Singh and AKS Rawat (2015). Comparative botanical and phytochemical evaluation of *Calotropis procera* Linn. and *C. gigantea* Linn. Root. J of Applied Pharma Sci 5(7): 041-047.

17

Coleus forskohlii Briq. Syn. *Plectranthus barbatus* Andr.

Family

Lamiaceae

Vernacular names [1]

Sanskrit: Makandi, Pashanbhed; Hindi: Patharchur; English: Coleus; Gujarati: Garmalu; Kannada: Makandiberu.

Distribution [2-3]

The species is native of Indian sub-continent, grows wildly in warm temperate climate of India, Nepal and Sri Lanka. It is also distributed in Egypt, Ethiopia and Brazil. The species is widely cultivated in Tamil Nadu, Karnataka, Maharashtra and Andhra Pradesh regions of India.

Traditional Uses [4, 6]

In Ayurveda it is used to treat hypertension, respiratory disorder, asthma, intestinal disorder and congestive heart failure. Kothas tribe of Nilgiri (India) uses decoction of root as a tonic. It is also used in cough, skin infections and root juice is used in constipation.

Flowering Plant

Fresh Roots

Major Chemical Constituents [4-5]

Forskolin and Iso-forskolin

Major Pharmacological Activity [7-10]

Anti-thrombotic, anti-obesity, anti-hypertensive, anti-aging and in alzheimer's disease.

Plant Description

Coleus forskholii Briq. is an evergreen, perennial plant, grows up to 1 meter in height, quadrangular stem with tuberous root. Leaves are small, ovate or oblong, pubescent, green, fleshy and petiolate. Stem are woody, erect, hard, quadrangular, branched, divided into nodes and internodes and nodes are hairy. Inflorescence is raceme; Flowers are stout, small and violet in colour. Upper calyx is broadly ovate and hairy inside, corolla is bilabiate. The roots are pale or golden brown coloured, tuberous with few rootlets, wrinkled, characteristic aromatic odour and pungent taste.

Anatomy

Transverse section of mature root is circular in outline. The outer layer is made up of multilayered rectangular cork cells, consists 1-2 layered cork cambium; epidermis is not distinguished due to secondary growth. The cortical region is made up of multilayered irregular parenchymatous cells; inner cortical cells are broad having intercellular spaces. Just below the cortical region phloem having sieve tubes, companion cells and phloem parenchyma are present, Cambium is well distinguished. Medullary rays are well developed, thin walled, radiating, heterogeneous. The xylem consists of vessels, fibers and tracheids.

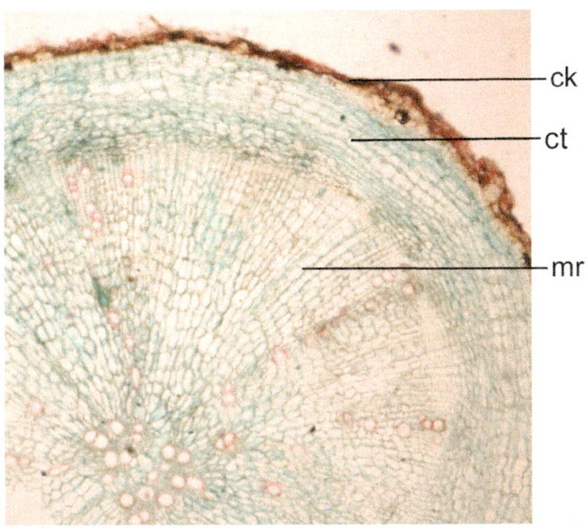

T.S. (4 x)

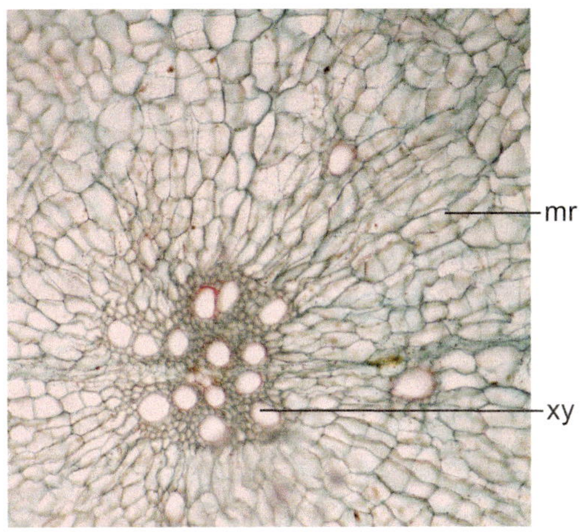

T.S. (10 x)

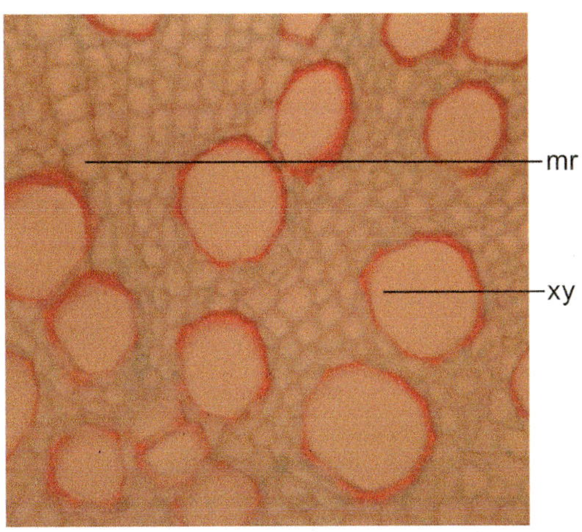

T.S. (40 x)

Transverse Section (T.S.) of *Coleus forskohlii* Root.

Abbreviation: ck- cork; ct- cortex; mr- medullary rays; xy- xylem.

HPTLC Studies

Sample Solution

Accurately weighed (5 g) crude powdered sample was subjected to cold extraction with methanol (50 ml), solution was continuously shaken for 6 h and then allowed to stand at room temperature (25 °C ± 2) for 18 hrs. The extraction process was repeated thrice, filtered and the pooled filtrate was concentrate in rotatory evaporator (Buchi, Switzerland). Working solution of 10 mg/ml was freshly prepared on the day of quantative analysis.

Standard Solution

Forskolin (≥98 per cent) and Iso-forskolin (≥98 per cent) was freshly prepared of strength 1 mg/ml.

Sample Application

2 µl of standard solution and 10 µl of sample was applied as 6 mm band with CAMAG HPTLC applicator on activated silica gel 60 GF_{254} HPTLC plate. The chromatogram was developed on pre-saturated CAMAG HPTLC twin trough chamber for 20 min and the solvent system was allowed to run up to 8 cm.

Solvent System

Toluene: ethyl acetate: methanol (9:3:0.05 v/v)

Visualization

The chromatogram was visualized after derivatization with anisaldehyde sulphuric acid reagent and densitometric scanning of forskolin and iso-forskolin was done at 545 nm.

Identification of Forskolin and Iso-forskolin[11-16]

The forskolin and iso-forskolin were identified at R_f 0.24 and 0.44 respectively.

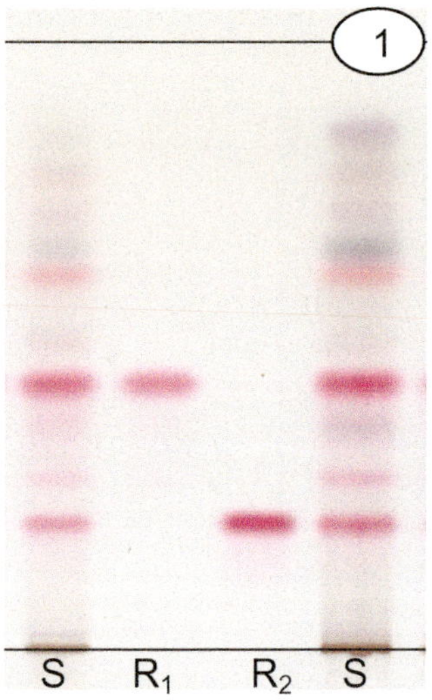

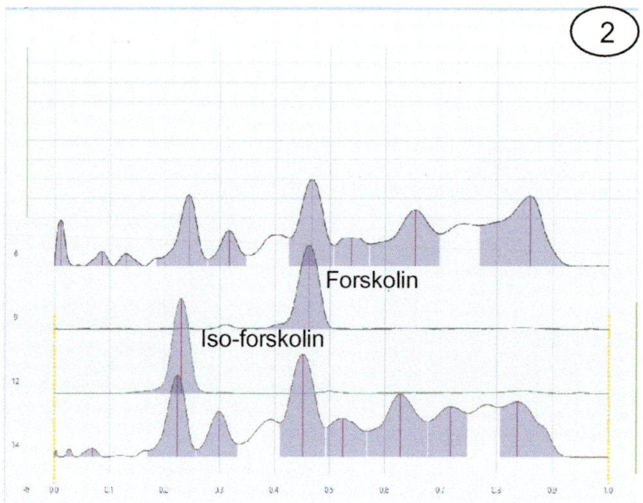

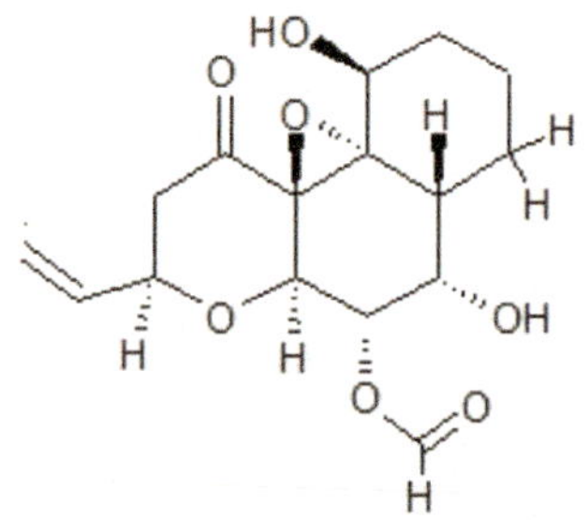

Forskolin

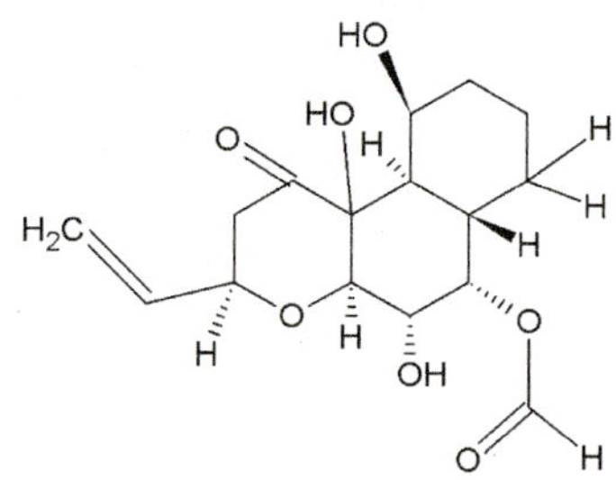

Iso-forskolin

Plate 1: HPTLC Chromatogram of *Coleus forskohlii* Sample (S) and Standard Forskolin (R_1) and Iso-forskolin (R_2).
Plate 2: Overlay Spectra of Sample and Standard.

1. Handa SS, Kaul MK (1996). Supplements to Cultivation and Utilization of Medicinal Plants. Regional Research Laboratory, CSIR, Jammu-Tawi.
2. Kurian A, Sankar A (2007). Other Important medicinal plants in trade. New India Publishing Agency, New Delhi, Edition1; 2: 291-320.
3. Willemse RH (1985). Notes on East African *Plectranthus* species (Lamiaceae). Kew Bull 40: 93-96.
4. Ammon HPT, Muller AB (1985). Forskolin: from an ayurvedic remedy to a modern agent. Planta Medica 46: 473-477.
5. Dubey MP, Srimal RC, Nityanand S, Dhawan BN (1981). Pharmacological studies on Coleonol, a hypotensive diterpene from *Coleus forskohlii*. J Ethnopharmacol 3: 1-13.
6. Jain SK (1981). Glimpses of Indian Ethanobotany. Oxford and IBH Publishing Co., Bombay.
7. De Souza NJ, Dohadwalla AN, Reden J (1983). Forskolin a labdane diterpenoid with antihypertensive, positive iontropic, platelet aggregation inhibitory, and adenylate cyclase activating properties. Med Res 3: 201-19.

8. Han LK, Morimoto C, Yu RH, Okuda H (2005). Effects of *Coleus forskohlii* on fat storage in ovariectomized rats. Yakugaku Zasshi 125: 449-453.
9. Siegl AM, Daly JW, Smith JB (1982). Inhibition of aggregation and stimulation of cyclic AMP generation in intact human platelets by the diterpene forskolin. Mol Pharmacol 21:680-687.
10. Dubey MP, Srimal RC, Nityanand S, Dhawan BN (1981). Pharmacological studies on coleonol, a hypotensive diterpene from *Coleus forskohlii*. J Ethnopharmacol 3: 1-13.
11. Pushpendra Kumar Shukla, Ankita Misra, Manish Kumar, Jaichand, Kuldeep Singh, Juber Akhtar, Sharad Srivastava, P K Agrawal and AKS Rawat (2018). Simultaneous Quantification of Forskolin and Iso forskolin in *Coleus forskohlii* (Briq.) and Identification of Elite Chemotype from Eastern Ghats (India). Pharmacog Mag 13: 881-885.
12. Sharad Srivastava, Ankita Misra, Priyanka Mishra, Pushpendra Shukla, Manish Kumar, V Sundaresan, KS Negi, Pawan K Agarwal, and AKS Rawat (2017). Molecular and chemotypic variability of forskolin in *Coleus forskohlii* Briq., a high value industrial crop collected from Western Himalayas (India). RSC Adv., 7, 8843–8851.
13. Pushpendra Kumar Shukla, Ankita Misra, Sneh Lata, S Rajan, Sharad Srivastava and AKS Rawat (2016). Intra-specific chemotypic variability of forskolin content in *Coleus forskohlii* (Wild.) Briq. growing in Nilgiri hills of India. J Planar Chromto 29(5): 347-355.
14. Ankita Misra, Sharad Srivastava, Pankaj Srivastava, Pushpendra Shukla, Pawan K Agrawal and AKS Rawat (2016). Chemotaxonomic variation in forskolin content and its correlation with ecogeographical factors in natural habitat of *Coleus forskohlii* Briq. collected from Vidarbha (Maharashtra, India). Ind Crops Prod, 84: 50-58.
15. Sharad Srivastava, Shikhar Verma, Abhishek Gupta, Pradeep Kushwaha, and AKS Rawat (2014). Chemotaxonomic studies of Forskolin in *Coleus forskohlii* Briq. from Different Phyto-Geographical Zones of India. In P.K. Mukherjee ed. Traditional Medicine and Globalization: The Future of Ancient Systems of Medicine (Maven Publishers) 322-330.
16. Sharad Kumar Srivastava, Manjoosha Chaubey, Sayyada Khatoon, A.K.S. Rawat and S. Mehrotra (2002). Pharmacognostic studies of *Coleus forskohlii* Briq. Pharmaceutical Biol. 40 (2): 129-134.

18

Coptis teeta Wall.

Family

Ranunculceae

Vernacular name

Sanskrit: Supita; Hindi: Mamira; Assamese: Mishmi tita; English: Gold thread; Tamil: Pitarohini.

Distribution [1]

The plant is endemic to Arunachal Pradesh, however it is also found in Meghalaya (India), China, Nepal and Tibet.

Traditional Uses [2]

In Arunachal Pradesh, tribes use *C. teeta* in malaria, abdominal pain, dysentery, cold, diarrhea, typhoid, high blood pressure, Jaundice and in eyes conjunctivitis.

Major Chemical Constituents [3-4]

Berberine, coptisine, jatrorrhizine, palmatine.

Major Pharmacological Activity [5-6]

Anti-tumor, anti-oxidant and anti-inflammatory activity.

Plant Description

It is a small perennial evergreen herb, approximately 1 foot high and stem are absent. Leaves are dark green, glabrous, petioles are long, pinnate, lamina 3 lobed and

Species in Natural Location
***Courtesy*: Arunachal Pradesh State Medicinal Plant Board.**

Fresh Root
***Courtesy*: http://dentomed.toyama-wakan.net/en/scientific_information_on_crude_drugs/Coptis%20Rhizome/SCC000007?w=1).**

prominent lateral veins are clearly seen. Flowers small, white or yellow and seeds are black. Rhizome irregular, leaf scars are prominent and give rough appearance, yellowish brown appearances with too many fibrous roots are arises from rhizome.

Anatomy

The transverse section of root is almost circular in outline shows narrow range of cork and central wood, encircled by narrow phloem and wider cortex. The outermost zone is made of cork cell which are rectangular and multilayered. Cortex is wider, parenchymatous and contains berberine alkaloid in the cells. Stele are well developed and arranged in a ring, xylem vessels are solitary or in group of 2 to 5. Medullary rays are homogenous and running along with xylem vessels and centrally parenchymatous pith are present.

HPTLC Studies

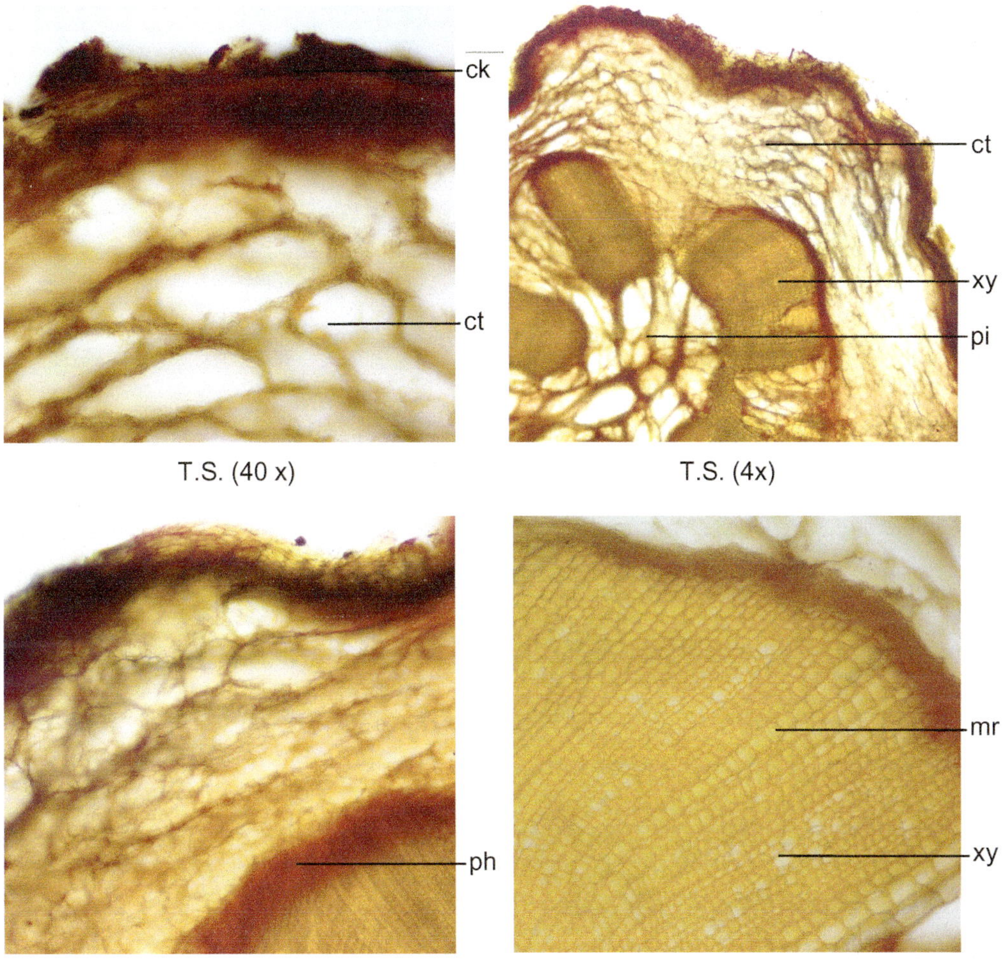

Transverse Section (T.S.) of *Coptis teeta* Rhizome.

Abbreviation: ck- cork; ct- cortex; mr- medullary rays; mx- metaxylem; ph- phloem; pi- pith; xy- xylem.

Sample Solution

Accurately weight (5 g) crude powdered rhizome was subjected to cold extraction with methanol (50 ml), continuously shaken for 6 h and then allowed to stand at room temperature (25 °C ± 2) for 18 h. The extraction was repeated thrice, filtered and the pooled filtrate was concentrate in rotatory evaporator (Buchi, Switzerland). Working solution of strength 10 mg/ml was freshly prepared on the day of quantative analysis.

Standard Solution

Berberine (≥ 98 per cent) was freshly prepared of strength 1 mg/ml

Sample Application

2 μl of standard solution and 10 μl of samples was applied as 6 mm band with CAMAG HPTLC applicator on activated silica gel 60 GF_{254} HPTLC plate. The chromatogram was developed on pre-saturated CAMAG

HPTLC twin trough chamber for 30 min and the solvent system was allowed to run for 8 cm.

Solvent System

n-propanol: water: formic acid (9:1:0.1 v/v)

Visualization

The chromatogram was visualized at UV_{366} and densitometric scanning of berberine was done at 350 nm.

Identification of Berberine

The standard berberine was identified at R_f 0.32

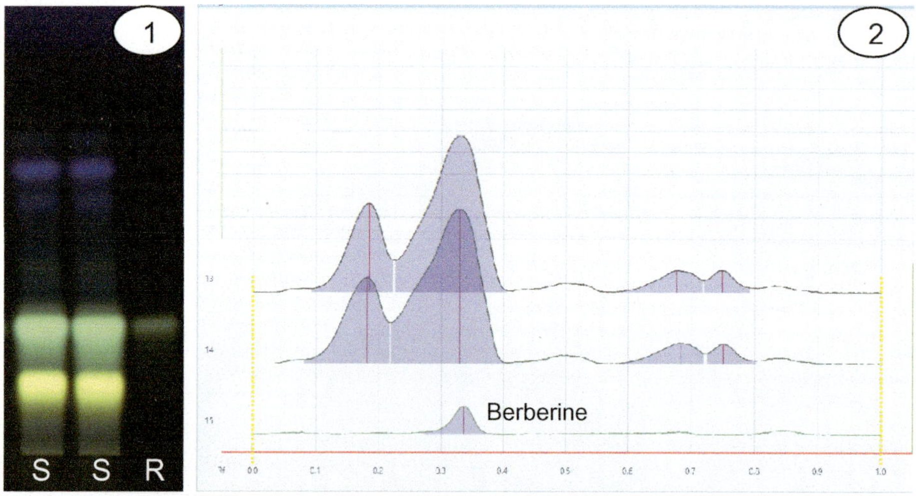

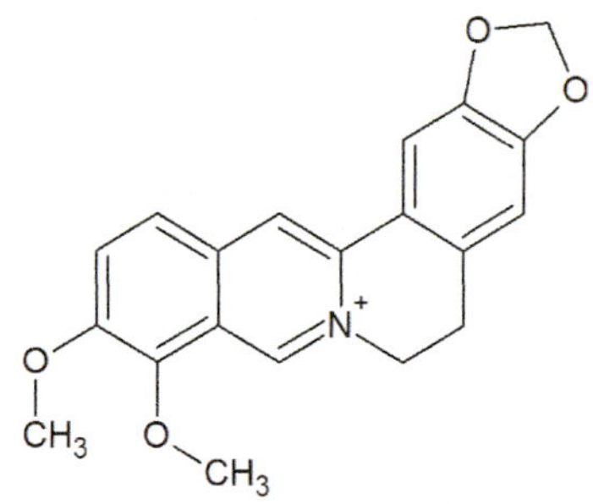

Berberine

Plate 1: HPTLC Chromatogram of *Coptis teeta* Sample (S) and Standard Berberine (R).
Plate 2: Overlay Spectra of Sample and Standard.

1. Selvam AB (2012). Pharmacognosy of negative listed Plants. Botanical Survey of India, Ministry of Environment and Forests.
2. Kanjilal, Borthakur (1997). Flora of Assam: (1997 reprint); Omsons Publications, New Delhi.1: 9. 1100270.
3. Li CY, Tsai SI, Damu AG, Wu TS (2009). A rapid and simple determination of protoberberine alkaloids in *Rhizoma coptidis* by 1H NMR and its application for quality control of commercial prescriptions; J Pharma Biomed Ana 49: 1272-1276.
4. Payum T (2017). Distribution, Ethnobotany, Pharmacognosy and Phytoconstituents of *Coptis teeta* Wall.: A Highly Valued and Threatened Medicinal Plant of Eastern Himalayas. Pharmaco J 9(6s).
5. Jeong JB, Jeong HJ (2010). 2-Methoxy-4-vinylphenol can induce cell cycle arrest by blocking the hyper-phosphorylation of retinoblastoma protein in benzo[a]pyrenetreated NIH3T3 cells. Biochem Biophys Res Communi: 752-7.
6. Jeong JB, Hong SC, Jeong HJ, Koo JS (2011). Anti-inflammatory Effect of 2-Methoxy4-Vinylphenol via the Suppression of NF-κB and MAPK Activation, and Acetylation of Histone H3. Arch Pharm Res 34(12):2109-16.

19

Costus speciosus (Koen. ex Retz.)

Family

Costaceae

Vernacular names

Sanskrit: Kembuka, Kebuka; Hindi: Keu, Keukand; Bengali: Keumut; Marathi: Penva, Pinga; English: Spiral flag; Gujarati: Valakdi; Telengu: Cengalvakostu; Malayalam: Cannakkuvva.

Distribution [1]

Costus speciosus (Koen. ex Retz.) is native to the Malay Peninsula of South East Asia. The species grows widely in India ranging from Central India to Sub-Himalayas, Western Ghats, Maharashtra, Karnataka and Kerala.

Traditional Uses [2]

Rhizome is used in rheumatism, dropsy, pneumonia and juice of rhizome is applied to head for cooling and relief from head-ache. The rhizome is also used as aphrodisiac, purgative, anthelminthic, depurative, febrifuge, expectorant, tonic and improves digestion.

Major Chemical Constituents [3–5]

Diosgenin, dioscin, gracillin and curcumin.

Flowering Plant

Fresh Rhizome

Major Pharmacological Activity [6–9]

Anti-diabetic, anti-bacterial, anti-oxidant and larvicidal activity.

Morphological Description

It is a tropical herbaceous, succulent perennial herb, grows up to 2 m height with tuberous stem, sub-woody at base and horizontal rhizomes. Leaves are simple, smooth, sub-sessile, elliptic or ovate, spirally arranged around the stem. Flowers are large, white, terminal spike inflorescence with bright red bracts and black seed. The fresh rhizome is curved or sometime straight, cylindrical, 12-25 cm long and 2-6 cm in width. There are many rootlets arise from entire surface of rhizome.

Anatomy

Transverse section of rhizome is circular in outline, shows single layer of epidermis with wider cortex occupying major area of section, well distinguished endodermal layer and wide stellar system in ground tissue. Detailed structure accounted that the rhizome shows single layer epidermis made of 1 or 2 row, followed by multi-layered thin-walled cork cells; cortex and wide central ground tissue is parenchymatous and stealer vascular bundle is scattered throughout ground tissue. Steles are collateral, non-lignified fibro-vascular bundles; they have been more numerous adjacent peripheral region of endodermis. Stellar vascular bundles are well developed and composed of 2 to 5 vessels associated with narrow phloem. Simple oval to pear shaped starch grains are present throughout the ground tissue. In longitudinal section vessels are seen vertically running broad segmented canals, xylem and phloem parenchyma are seen as narrow elongated tubes, xylem fibres and phloem fibres are vertically arranged.

HPTLC Studies

Sample Solution

Accurately weighed (5 g) crude powdered rhizome was hydrolyzed with (50 ml) of 2.5 N hydrochloric acid

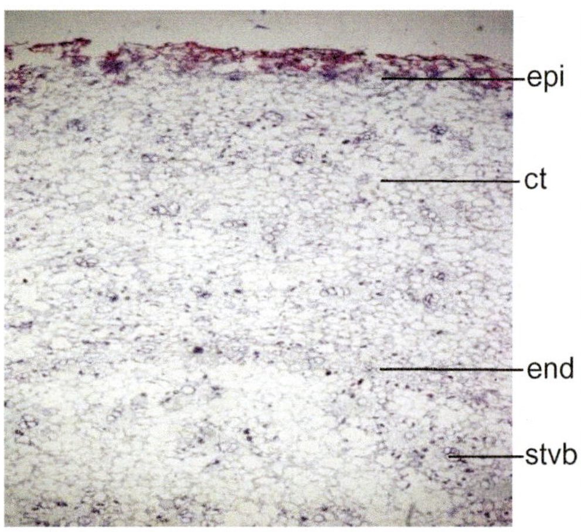

T.S. (4 x)

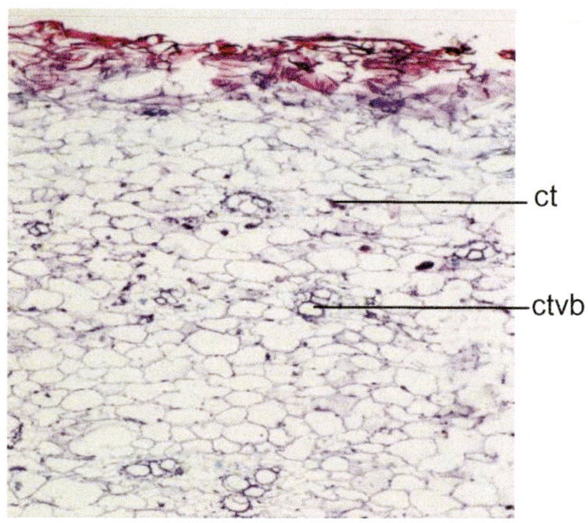

T.S. (20 x)

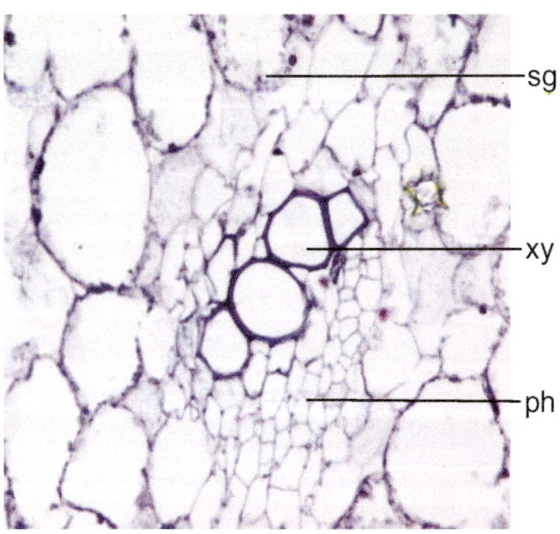

T.S. (40 x)

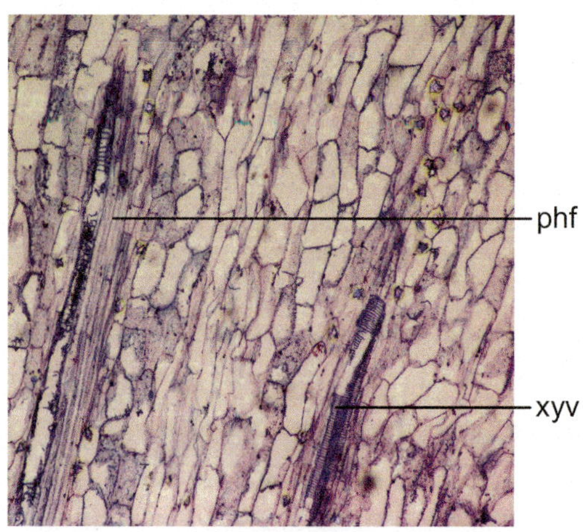

L.S. (20 x)

Transverse Section (T.S.) & Longitudinal Section (L.S) of *Costus speciosus* Rhizome.

Abbreviation: ct- cortex; ctvb- cortical vascular bundle; end- endodermis; epi- epidermis; ph- phloem; phf- phloem fibre; sg- starch grains; vb- vascular bundle; xy- xylem; xyv-xylem vessels.

for 4 hrs at 80 °C. The residue was filtered, washed with hot water until neutral or free from acid. The residue was then dried in hot air oven and further extracted with methanol (50 ml) by continuously shaken for 6 h and then allowed to stand for 18 h at room temperature (25 °C ± 2). Extraction was repeated thrice, filtered and the pooled filtrate was concentrate in rotatory evaporator (Buchi, Switzerland), dried residue was stored at 4 °C. Working solution of 10 mg/ml was freshly prepared on the day of analysis.

Standard Solution

Diosgenin (≥ 98 per cent) was freshly prepared of strength 1 mg/ml.

Sample Application

2 µl of standard solution and 10 µl of sample was applied as 6 mm band with CAMAG HPTLC applicator on activated silica gel 60 GF_{254} HPTLC plate. The chromatogram was developed on pre-saturated CAMAG HPTLC twin trough chamber for 20 min and the solvent system was allowed to run up to 8 cm.

Solvent System

Hexane: ethyl acetate (7:2 v/v)

Visualization

The chromatogram was visualized after derivatization with anisaldehyde sulphuric acid reagent and densitometric scanning of diosgenin was done at 440 nm.

Identification of Diosgenin[10-12]

The standard diosgenin was identified at R_f 0.31.

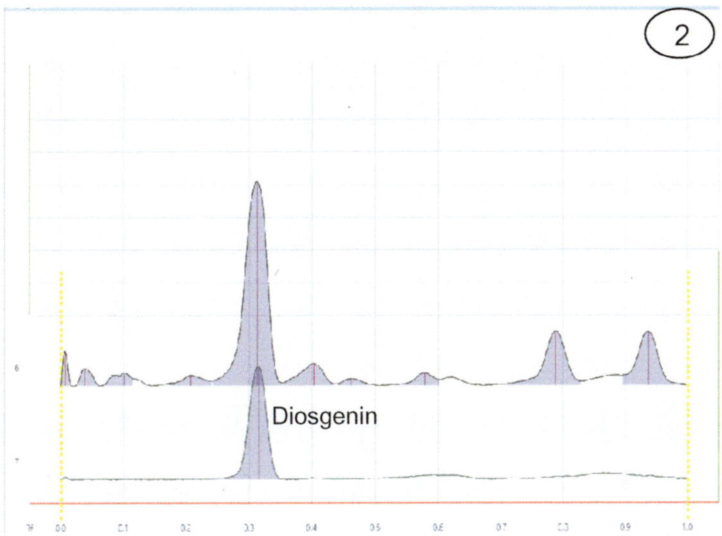

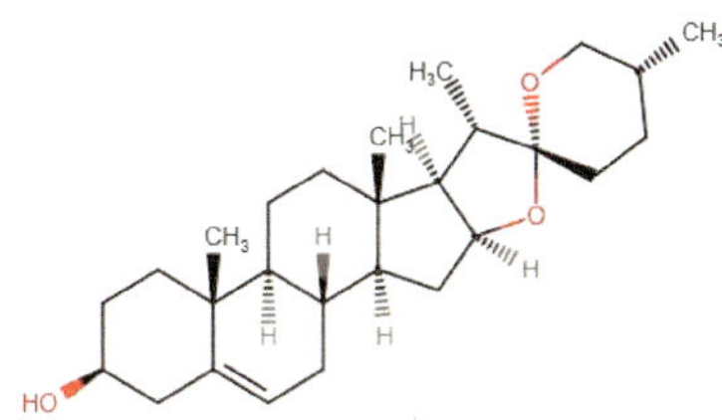

Diosgenin

Plate 1: HPTLC Chromatogram of *Costus speciosus* Sample (S) and Standard Diosgenin (R).
Plate 2: Overlay Spectra of Sample and Standard.

1. Sarin YK, Bedi KL, Atal CK (1974). *Costus speciosus* rhizome as source of diosgenin. Curr sci 43(18): 569-570.
2. Anonymous (2007). The Wealth of India: First Supplement Series (Raw Materials), National Institute of Science Communication and Information Resources. CSIR Vol 2: 211-213.
3. Dasgupta B, Pandey VB (1970). A new Indian source of diosgenin (*Costus speciosus*). Cellu Mol Life Sci. 26(5): 475-6.
4. Rani SA, Sulakshana G, Patnaik S (2012). *Costus speciosus*, An Antidiabitic plant - Review. FSJ Pharma Res 1(3): 52-53.
5. Gupta AK, Tondon N, Sharma M (2008). Quality Standards of Indian Medicinal Plants, Medicinal Plants Unit, Published by Indian Council of Medical Research, Vol. VII, 48.
6. Rajesh MS, Harish MS, Sathyaprakash RJ, Shetty AR, Shivananda TN (2009). Antihyperglycemic activity of the various extracts of *Costus speciosus* rhizomes. J Nat Remedies 9(2): 235-241.
7. Malabadi RB (2005). Antibacterial activity in the rhizome extracts of *Costus speciosus* (Koen.). J Phytol Res 18(1): 83-5.
8. Nehete J, Bhatia M, Narkhede M (2010). In-vitro Evaluation of Antioxidant Activity and Phenolic

Content of *Costus speciosus* (Koen) J. E. Sm. Iran J Pharm Res 9(3): 271-277.

9. Kanakkanath S, Narayanan A, Ganesan B (2013). Evaluation of *Costus speciosus* (Koen.) aqueous extract for larvicidal activity. Der Pharmacia Lettre 5(4): 283-285.

10. Poonam Rawat, Manish Kumar, Akanksha Srivastava, Bhanu Kumar, Ankita Misra, Satyendra Pratap Singh, Sharad Srivastava (2021). Diosgenin content mediated diversity assessment of *Costus speciosus* (Koen. Ex Retz) collected from Indo-Gangetic plains and their correlation with soil dynamics. Chem Biodivers, DOI: 10.1002/cbdv.202000977.

11. Manish Kumar, Ankita Misra, Akanksha Srivastava, Pushpendra Kumar Shukla, Lalit Mohan Tewari, Sharad Srivastava (2020). Comparative Pharmacognostical and Pharmacological Evaluation of *Costus speciosus* (Koen) J.E. Sm. Germplasm Collected from Eastern Ghats of India. Pharmacog J 12(1):150-156. DOI: 10.5530/pj.2020.12.

12. Akanksha Srivastava, Ankita Misra, Manish Kumar, Pushpendra Kumar Shukla, Pawan K Agarwal, Sharad Srivastava (2019). HPTLC densitometric quantification of Diosgenin content in *Costus speciosus* germplasm collected from Eastern Ghats of India and identification of elite chemotypes. Pharmacog Mag 15(66): 462-467.

20

Curcuma amada Roxb.

Family

Zingiberaceae

Vernacular name [1]

Sanskrit: Amrardrakam; Hindi: Ama-haldi; English: Mango ginger; Gujarati: Valakdi; Telugu: Mamidi Allam; Malayalam: Mangainchi; Tamil: Mankayinci.

Distribution [2-3]

C. amada is wildely available in West Bengal and is cultivated in Gujarat, Uttar Pradesh, Kerala, Karnataka, Tamil Nadu and North-Eastern states of India.

Traditional Uses [4-5]

Curcuma amada is medically used as a coolant, aromatic, astringent and promotes digestion. The paste of rhizome has traditionally been used for healing wounds, cuts and in itching. The species have raw mango-like aroma and is used in making salad and, pickles in different parts of India.

Major Chemical Constituents [6-7]

Curcumin, de-methoxy curcumin and bis-demethoxy curcumin.

Pharmacological Activity [8-10]

Anti-microbial, anti-oxidant and anti-inflammatory activity.

A Flowering Stalk with Rhizome

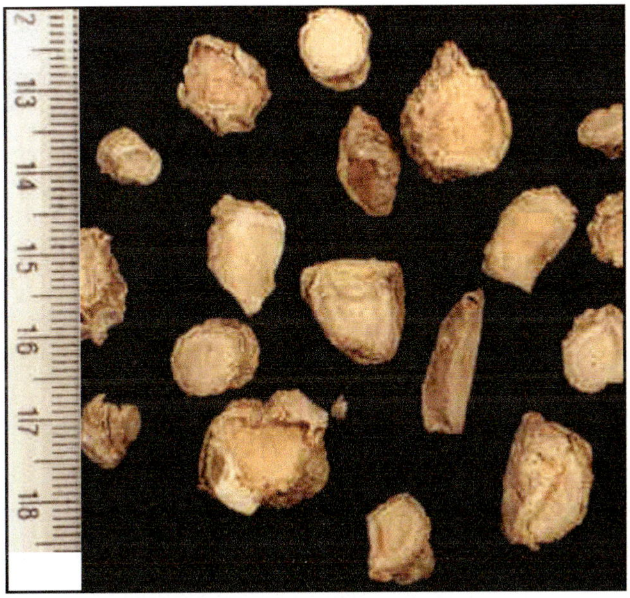

Dried Pieces of Rhizome

Morphological Description

Curcuma amada Roxb. is commonly known as mango ginger. It is a herbaceous perennial plant, erect or semi-erect, rhizomatous, aromatic herb grows up to 1 meter. The leaves are long, petiolate, radical and oblong-lanceolate shaped with pointed leaf apex. The rhizome is large and branched, fleshy and divided into nodes and internodes with a buff-colored external surface. At nodes, scaly leaves are arranged in circular giving the appearance of growth rings with scars on the surface. The rhizomes imitate a raw mango flavor and pungent taste.

Anatomy

Transverse section of rhizome is almost circular in outline, show very narrow range of cork and wider cortex occupying major area of section, well distinguished endodermal layer and wide stellar system in ground tissue. Detailed structure accounted that the rhizome shows single layer of epidermis followed by multi-layered thin-walled cork cells lie underneath this; cortex and wide central ground tissue is parenchymatous and stealer vascular bundle is scattered throughout ground tissue. Well defined single layer endodermis is present. Steles are collateral, non-lignified fibro-vascular bundles; they have been more numerous adjacent to peripheral region of endodermis. Oleoresin canals are often seen throughout the tissues. Stellar vascular bundles are well developed and composed of 2 to 10 vessels associated with narrow phloem. Oleoresin canals and simple oval to pear shaped starch grains are present throughout the ground tissue as well as in cortex region. In longitudinal section vessels are seen vertically running broad segmented canals, xylem and phloem parenchyma are seen as narrow elongated tubes, xylem vessels and phloem fibers are vertically arranged.

HPTLC Studies

Sample Solution

Accurately weight (5 g) crude powdered sample was subjected to cold extraction with methanol (50 ml), continuously shaken for 6 h and then allowed to stand at room temperature (25 °C ± 2) for 18 h. The extraction

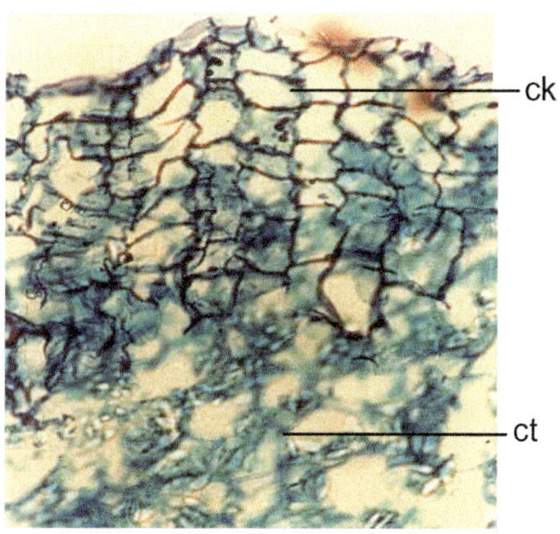

T.S. (40 x)

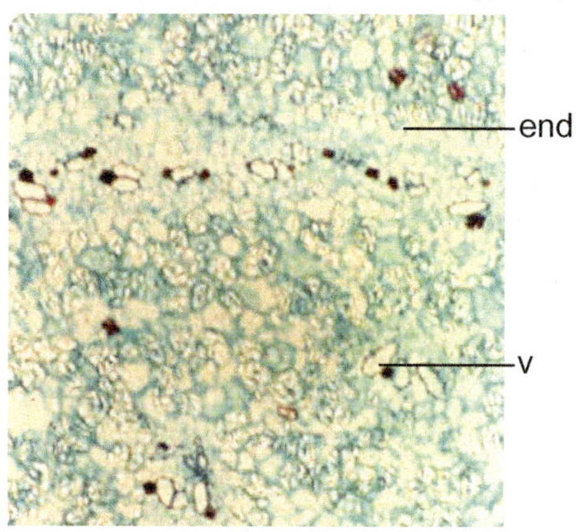

T.S. (10 x)

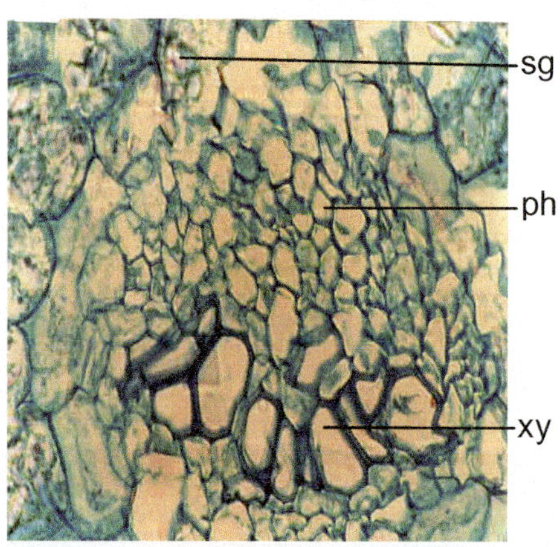

T.S. (40 x)

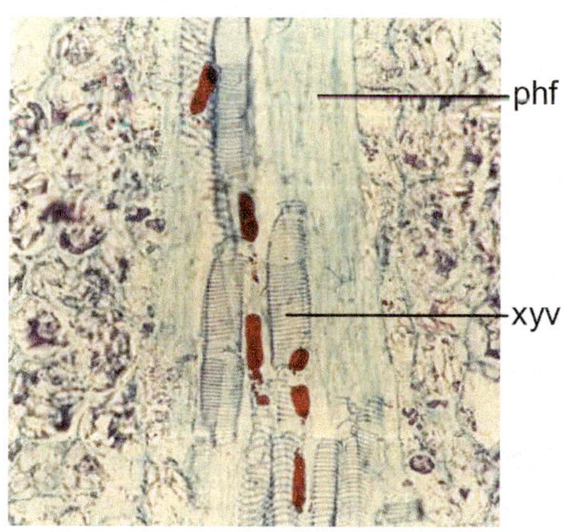

L.S. (20 x)

Transverse Section (T.S.) & Longitudinal Section (L.S) of *Curcuma amada* Rhizome.
Abbreviation: ck- cork; ct- cortex; ph- phloem; sg- starch grains; vb- vascular bundle; xy- xylem; xyv- xylem vessels.

was repeated thrice, filtered and the pooled filtrate was concentrate in rotatory evaporator (Buchi, Switzerland). Working solution of strength 10 mg/ml was freshly prepared on the day of quantative analysis.

Standard Solution

Curcumin (≥ 98 per cent) and demethoxy curcumin (≥ 98 per cent) was freshly prepared of strength 1 mg/ml.

Sample Application

2 µl of standard solution and 10 µl of sample was applied as 6 mm band with CAMAG HPTLC applicator on activated silica gel 60 GF_{254} HPTLC plate. The chromatogram was developed on pre-saturated CAMAG HPTLC twin trough chamber for 15 min and the solvent system was allowed to run for 8 cm.

Solvent System

Chloroform- ethanol- glacial acetic acid (9.4:0.5:0.1 v/v)

Visualization

The chromatogram was visualized at UV_{366} and densitometric scanning of curcumin and demethoxy curcumin was done at 420 nm.

Identification of Curcumin and Demethoxy Curcumin[11-13]

The standard demethoxy curcumin and curcumin was identified at R_f 0.37 and 0.46.

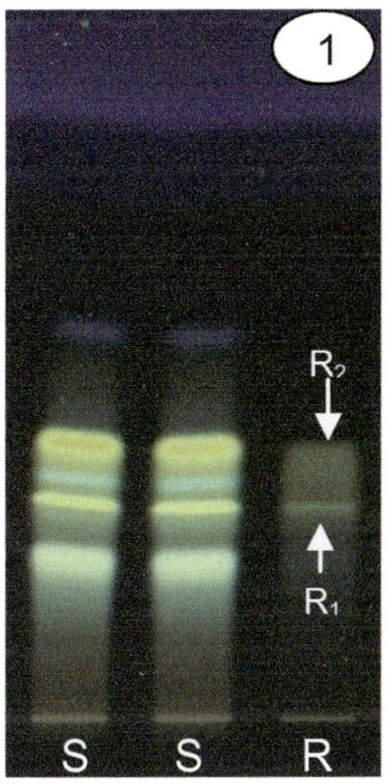

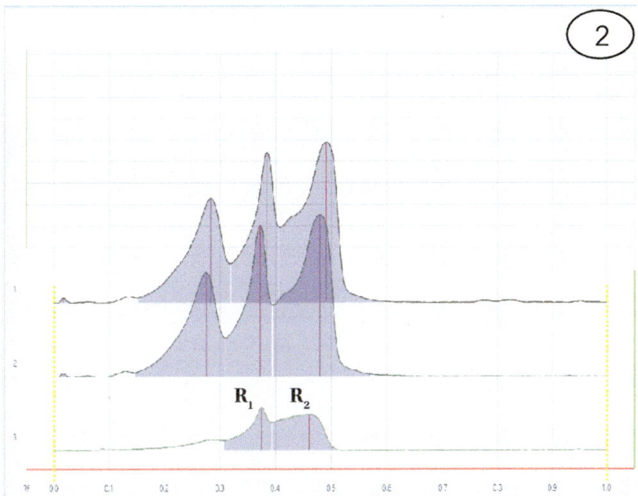

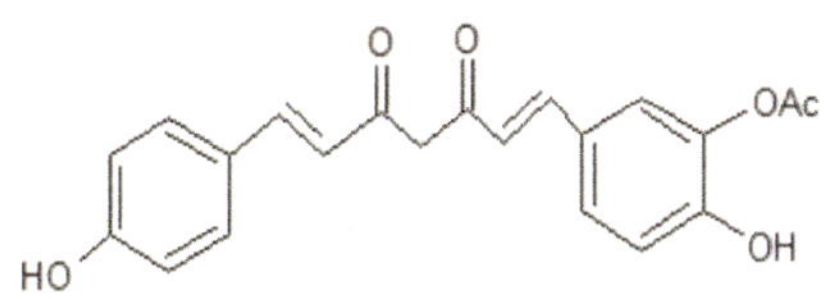

Demethoxy curcumin

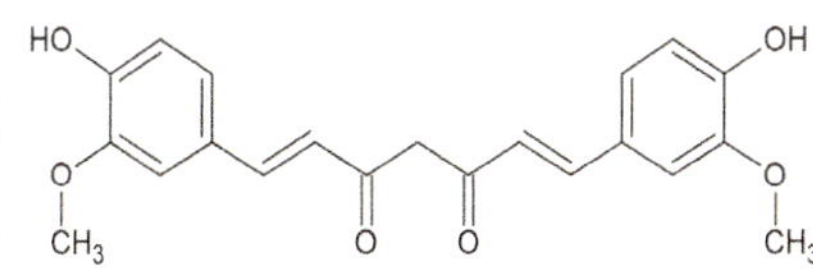

Curcumin

Plate 1: HPTLC Chromatogram of *Curcuma amada* (S) and Standard Demethoxy Curcumin (R1) and Curcumin (R2).
Plate 2: Overlay Spectra of Sample and Standard.

1. Samant LR (2012). *Curcuma amada* Roxb.: a phytopharmacological review. J Pharm Res 5(4): 1992-3.
2. Sasikumar B (2005). Genetic resources of *Curcuma*: diversity, characterization and utilization. Plant Gen Res 3: 230-251.
3. Velayudhan KC, Muralidharan VK, Amalraj VA, Gautam PL, Mandal S, Kumar D (1999). *Curcuma* genetic resources. Scientific monograph No. 4. National Bureau of Plant Genetic Resources Regional Station, Thrissur, Kerala, India: 1-27.
4. Srivastava S, Srivastava M, Rawat AKS, Mehrotra S (2006). Pharmacognostic evaluation of *Curcuma amada* Roxb. Proc Natl Acad Sci India 76: 153-160.
5. Hussain A, Virmani OP, Popli SP, Misra LN, Gupta MM (1992). Dictionary of Indian Medicinal Plants. Central Institute of Medicinal and Aromatic Plants: Lucknow 2: 39.
6. Policegoudra RS, Divakar S, Aradhya SM (2007). Identification of difurocumenonol, a new antimicrobial compound from mango ginger (*Curcuma amada* Roxb.) rhizome. J appl micro: 102(6):1594-602.
7. Gupta AP, Gupta AM and Sushil Kumar K (1999). Simultaneous determination of curcuminoids in *curcuma* samples using high performance thin layer chromatography. J Liq Chromatog Technol 22: 1561-1569.
8. Gupta RK, Gaur H, Pati JK, Pandey A (2015). Evaluation of antioxidant potential of different extracts of mango ginger (*curcuma amada* roxb.) rhizome. Int J Pharma Sci Res: 6(9): 3986-9.
9. Mujumdar AM, Naik DG, Dandge CN, Puntambekar HM (2000). Anti-inflammatory activity of *Curcuma amada* Roxb. in albino rats. Indian J Pharmacol 32: 375–377.

10. Policegoudra RS, Aradhya SM, Singh L (2011). Mango ginger (*Curcuma amada* Roxb.)–A promising spice for phytochemicals and biological activities. J biosci 36(4): 739-48.

11. Sharad Srivastava, Manjoosha Srivastava, Shanta Mehrotra and AKS Rawat (2009). A HPTLC Method for Chemotaxonomic evaluation of some *Curcuma* Species and their Commercial Samples. J Sci Ind Res 68:876-880.

12. Sharad Kumar Srivastava, Ajay Kumar Singh Rawat, Shanta Mehrotra and P. Pushpangadan (2007). Botanical standardization of three commercially important *Curcuma* species. J Sci Ind Res 66: 450-456.

13. Sharad Srivastava, M. Srivastava, A. K. S. Rawat and S. Mehrotra (2006). Pharmacognostic evaluation of roots of *Curcuma amada* Roxb. Proc Nat Acad Sci 75 (B), II: 88-95.

21

Curcuma angustifolia Roxb.

Family

Zingiberaceae

Vernacular name [1-3]

Sanskrit: Tavaksheera, Payaksheera; Hindi: Tikhur, Thikora; English: East Indian arrow root; Marathi: Tavakira; Tamil: Artimavu; Malayalam: Koova; Telugu: Gaddalu.

Distribution [4]

C. angustifolia is found in the Himalayas ranging from Kumaon to North-East India and South East Asia at an altitude ranging from 900-1200 meter. It is also found in Maharashtra, Chhattisgarh, Bihar, Andhra Pradesh, West Bengal, Tamil Nadu and Kerala.

Traditional Uses [5-6]

Curcuma is medically used as a coolant, aromatic, astringent and is used to promote digestion. Rhizome of *C. angustifolia* Roxb. is used to cure peptic ulcer, dysentery, diarrhea and is often employed as a herbal tonic for patients suffering from tuberculosis. It is also use for the treatment cough and bronchitis.

Major Chemical Constituents [7]

Curcumin, de-methoxy curcumin and bis-demethoxy curcumin.

A Flowering Twig

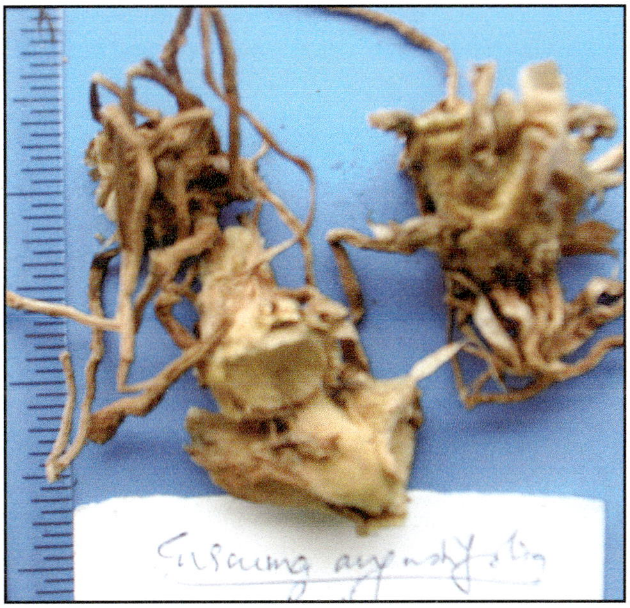

Dried Pieces of Rhizome

Pharmacological Activity [8–9]

Anti-proliferative and anti-oxidant activity

Morphological Description

Curcuma angustifolia Roxb. is commonly known as East Indian arrow root. It is an herbaceous perennial plant, erect or semi-erect, rhizomatous, aromatic herb. The leaves are simple, long, radical and oblong–lanceolate shaped, petiole may or may not present, glaberous or hairy with prominent mid-vein and numerous lateral veins. Inflorescence is terminal on pseudo-stem and raceme or spike type. Flowers are bisexual, epigynous, zygomorphic, calyx and corolla are tubular, stamens are 6 in number and ovary inferior. The rhizome is large and branched, fleshy and divided into nodes and internodes, with secondary root arises. At rhizome nodes scaly leaves are arranged circularly giving the appearance of growth rings with scars on the surface.

Anatomy

Transverse section of rhizome is almost circular in outline, show very narrow range of cork and wider cortex occupying major area of section, well distinguished endodermal layer and wide stellar system in ground tissue. Detailed structure accounted that the rhizome shows single layer of epidermis followed by multi-layered thin-walled cork cells lie underneath this; cortex and wide central ground tissue is parenchymatous and stealer vascular bundle is scattered throughout ground tissue. Well defined single layer endodermis is present. Steles are collateral, non-lignified fibro-vascular bundles; they have been more numerous adjacent to peripheral region of endodermis. Stellar vascular bundles are well developed and composed of 2 to 10 or more vessels associated with narrow phloem. Oleoresin canals and simple rounded to oval or sometime rod shaped starch grains are present throughout the ground tissue as well as in cortex region. In longitudinal section vessels are seen vertically running broad segmented canals, xylem and phloem parenchyma are seen as narrow elongated tubes, xylem vessels and phloem fibres are vertically arranged to each other.

HPTLC Studies

Sample Solution

Accurately weight (5 g) crude powdered sample was subjected to cold extraction with methanol (50 ml),

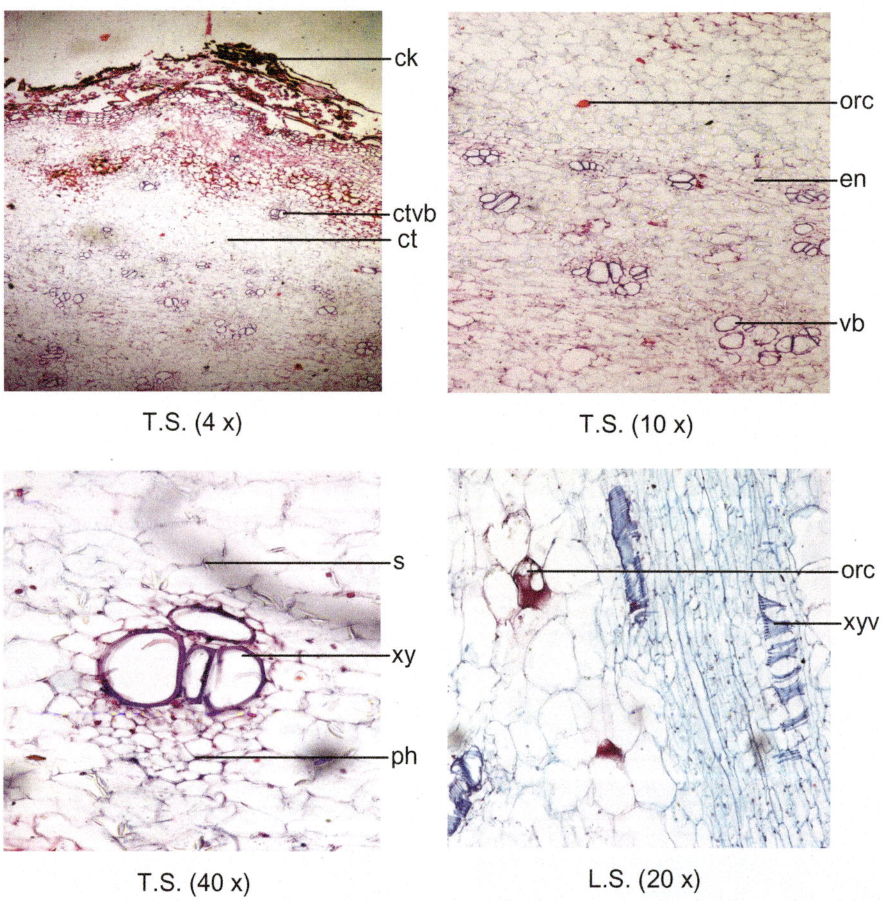

Transverse Section (T.S.) & Longitudinal Section (L.S) of *Curcuma angustifolia* Rhizome.

Abbreviation: ck- cork; ct- cortex; ctvb- cortical vascular bundle; end- endodermis; orc- oleoresin canals; ph- phloem; sg- starch grains; vb- vascular bundle; xy- xylem; xyv- xylem vessels.

continuously shaken for 6 h and then allowed to stand at room temperature (25 °C ± 2) for 18 h. The extraction was repeated thrice, filtered and the pooled filtrate was concentrate in rotatory evaporator (Buchi, Switzerland). Working solution of strength 10 mg/ml was freshly prepared on the day of quantative analysis.

Standard Solution

Curcumin (≥ 98 per cent), de-methoxy curcumin (≥ 95 per cent), bis-demethoxy curcumin (≥ 95 per cent) was freshly prepared of strength 1 mg/ml.

Sample Application

2 µl of standard solution and 10 µl of sample ws applied as 6 mm band with CAMAG HPTLC applicator on activated silica gel 60 GF_{254} HPTLC plate. The chromatogram was developed on pre-saturated CAMAG HPTLC twin trough chamber for 15 min and the solvent system was allowed to run for 8 cm.

Solvent System

Chloroform- ethanol- glacial acetic acid (9.4:0.5:0.1 v/v)

Visualization

The chromatogram was visualized at UV_{366} and densitometric scanning of Curcumin, de-methoxy curcumin and bis-demethoxy curcumin was done at 420 nm.

Identification of Reference Markers

The standard bis-demethoxy curcumin, de-methoxy curcumin and curcumin was identified at R_f 0.29, 0.39 and 0.49 respectively.

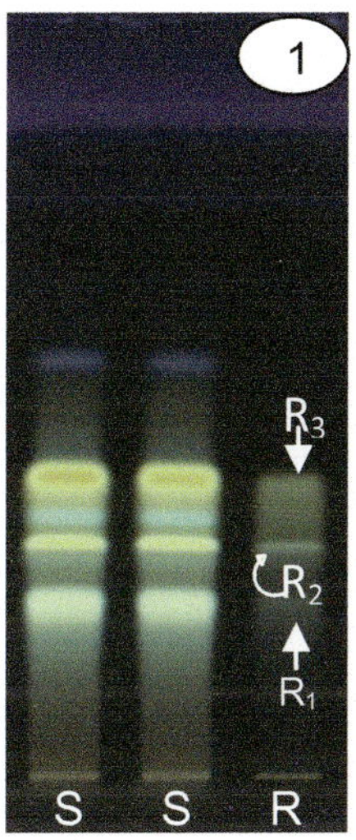

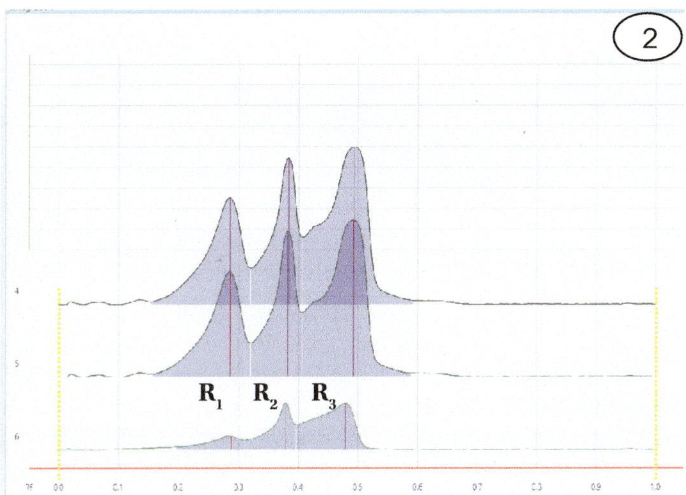

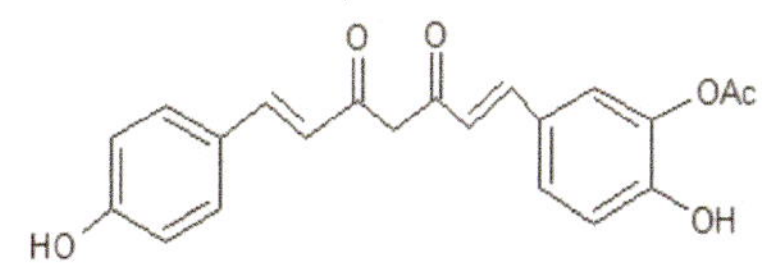

De-methoxy curcumin

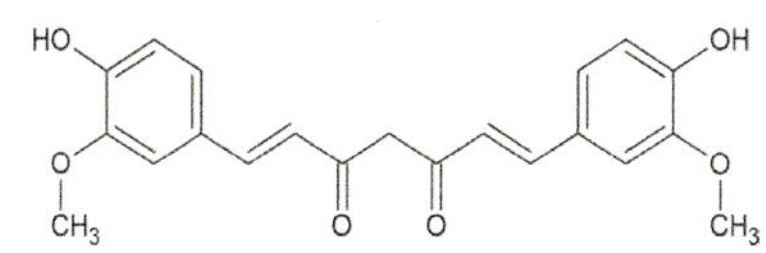

Curcumin

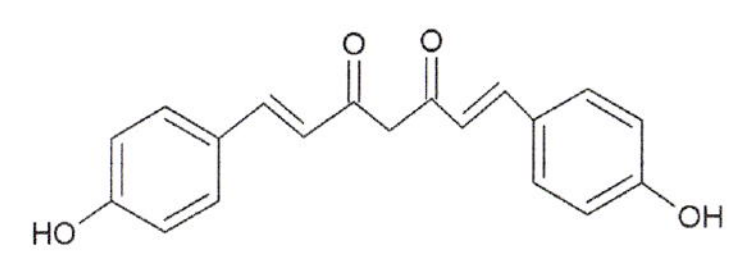

bis-demethoxy curcumin

Plate 1: HPTLC Chromatogram of *Curcuma angustifolia* (S) and Standard Bis-demethoxy Curcumin (R1), Demethoxy Curcumin (R2), Curcumin (R3).

Plate 2: Overlay Spectra of Sample and Standard.

1. Nadkarni AK (1976). Editor. Indian Materia Medica. Popular Prakashan Pvt. Ltd I: 413.
2. Indian medicinal plants. 2418.
3. Bapalal G (1985). nighantu Adarsha. Chaukhambha Bharati Academy, Varanasi.1st ed, 579.
4. Available from: http://www.flowersofindia.net/catalog/slides/East per cent 20 Indian per cent 20Arrowroot.html.
5. Tiwari S, Patel S (2013). A comparative study of tikhur traditional and partial mechanical processing and cost economics. Int J Agric Eng 6(1): 213-5.
6. Tiwari S, Patel S (2013). Physico-chemical variation on tikhur powder obtained from mother and finger rhizomes. J Plant Developm Sci 5(2): 139-41.
7. Srivastava S, Srivastava M, Rawat AKS, Mehrotra S (2006). Pharmacognostic evaluation of *Curcuma amada* Roxb. Proc Natl Acad Sci India 76: 153-160.
8. Doble B, Dwivedi S, Dubey K, Joshi H (2011). Pharmacognostical and antimicrobial activity of leaf of *Curcuma angustifolia* Roxb., Int J Drug Disco Herbal Res 1(2): 46-49.
9. Assumi SR, Puro KU, Kasomva K, Duarah A, Jha AK, Sen A (2017). In vitro anti-proliferative activity of *Curcuma angustifolia* and estimation of bioactive compounds for antioxidant activity. J Food Sci Eng 7: 59-66.

22

Curcuma aromatica Salisb.

Family

Zingiberaceae

Vernacular name

Sanskrit: Aranyaharidra, Vanaharidra; Hindi: Jangli haldi; English: Wild Turmeric.

Distribution [1-2]

C. aromatica Salisb. is widely available in South Asian countries and is distributed from China to Srilanka. In India, it is available in Himalayan region and Western Ghats. The species is also cultivated in southern parts of India.

Traditional Uses [3-4]

Curcuma is medically used as a coolant, aromatic, astringent and is digestive. It is traditionally used in India as a tonic, carminative and antidote to snake bites. Rhizome powder with milk is used in dysentery and gastric ailments. Rhizome paste is used in facial application to reduce acne, excessive hair growth and also used to improve skin tone and complexion.

Major Chemical Constituents [5-6]

Curcumin, de-methoxy curcumin, bis-demethoxy curcumin, limonine, camphene.

Pharmacological Activity [7-10]

Anti-inflammatory, anti-oxidant, anti-cancer and larvicidal activity.

Plant in Natural Habitat

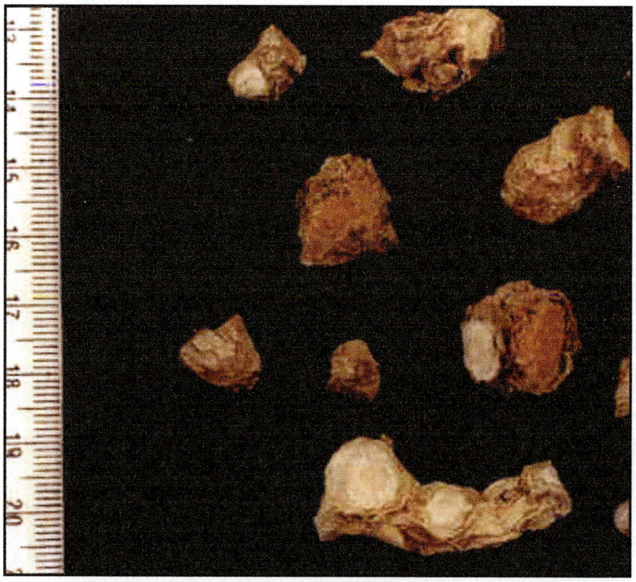

Dried Pieces of Rhizome

Morphological Description

Curcuma aromatica Salisb. is commonly known as Vanaharidra. It is an herbaceous perennial plant, erect or semi-erect, rhizomatous, aromatic herb, propagated by rhizome. The leaves are simple, long, elliptic and oblong–lanceolate shaped, glaberous or hairy with prominent mid-vein and numerous lateral vein. Inflorescence is terminal on pseudo-stem and raceme or spike type. Flowers are white or pink colored, bisexual, zygomorphic, calyx and corolla are tubular. Rhizomes have specific fragrance on attaining maturity. The rhizome is large and branched, fleshy and divided into nodes and internodes. At rhizome nodes scaly leaves are arranged circularly giving the appearance of growth rings with scars on the surface.

Anatomy

Transverse section of rhizome is almost circular in outline, show very narrow range of cork and wider cortex occupying major area of section and wide stellar system in ground tissue. Detailed structure accounted that the rhizome shows multi-layered thin-walled cork cells made up of 4-6 rows; cortex and wide central ground tissue is parenchymatous and stealer vascular bundle is scattered throughout ground tissue. Steles are collateral, non-lignified fibro-vascular bundles; they have been more numerous adjacent to peripheral region of endodermis. Oleoresin canals are very few in number in the tissues. Stellar vascular bundles are well developed and composed of 5 to 20 or more vessels associated with narrow phloem. Simple rod shaped starch grains are present throughout the ground tissue as well as in cortical region. In longitudinal section vessels are seen vertically running broad segmented canals, xylem and phloem parenchyma are seen as narrow elongated tubes, xylem vessels and phloem fibers are vertically arranged to each other.

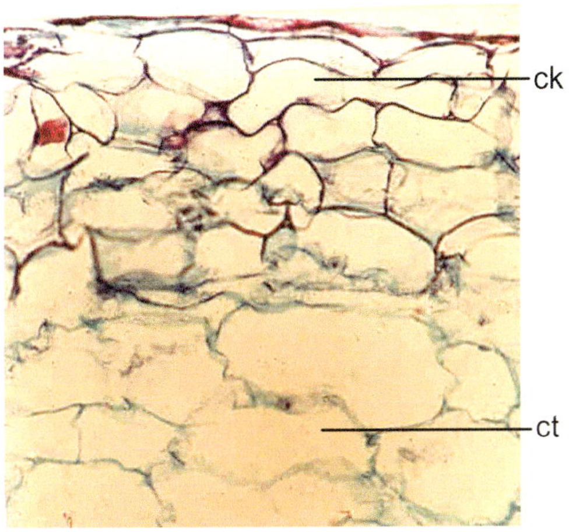

T.S. (40 x)

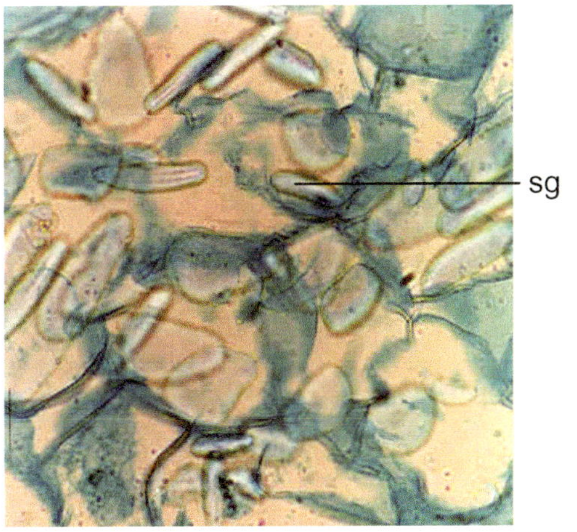

T.S. (40 x)

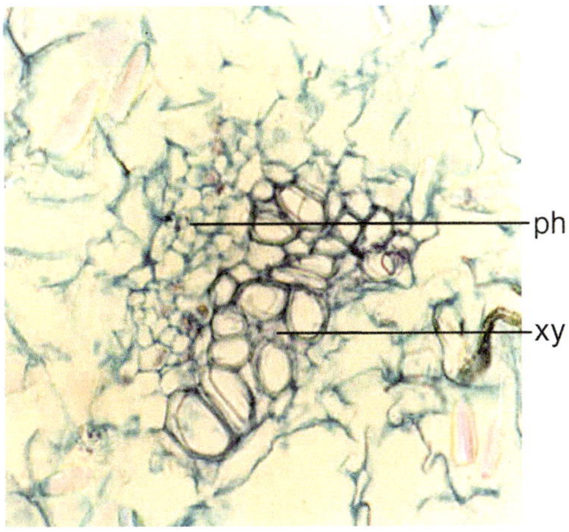

T.S. (20 x)

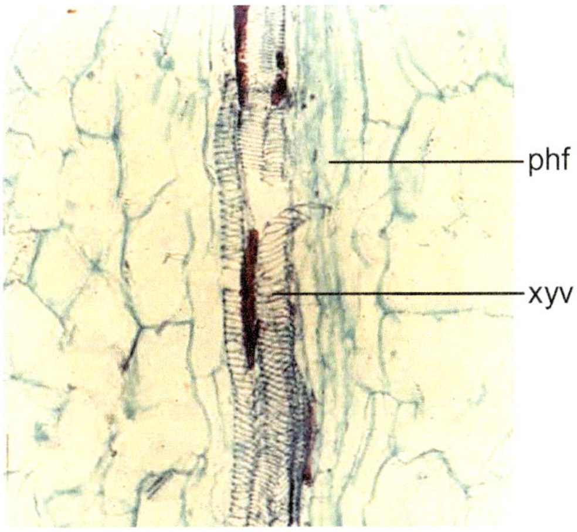

L.S. (20 x)

Transverse Section (T.S.) & Longitudinal Section (L.S) of *Curcuma aromatica* Rhizome.

Abbreviation: ck- cork; ct- cortex; ph- phloem; phf- phloem fibre; sg- starch grains; xy- xylem; xyv- xylem vessels.

HPTLC Studies

Sample Solution

Accurately weight (5 g) crude powdered sample was subjected to cold extraction with methanol (50 ml), continuously shaken for 6 h and then allowed to stand at room temperature (25 ^{0}C ± 2) for 18 h. The extraction was repeated thrice, filtered and the pooled filtrate was concentrate in rotatory evaporator (Buchi, Switzerland). Working solution of strength 10 mg/ml was freshly prepared on the day of quantative analysis.

Standard Solution

Curcumin (≥ 98 per cent), de-methoxy curcumin (≥ 98 per cent) and bis-demethoxy curcumin (≥ 98 per cent) was freshly prepared of strength 1 mg/ml.

Sample Application

2 µl of standard solution and 10 µl of sample wsa applied as 6 mm band with CAMAG HPTLC applicator on activated silica gel 60 GF_{254} HPTLC plate. The chromatogram was developed on pre-saturated CAMAG HPTLC twin trough chamber for 15 min and the solvent system was allowed to run for 8 cm.

Solvent System

Chloroform- ethanol- glacial acetic acid (9.4:0.5:0.1 v/v)

Visualization

The chromatogram was visualized at UV_{366} and densitometric scanning of bis-demethoxy curcumin, de-methoxy curcumin and curcumin, and was done at 420 nm.

Identification of Reference Markers[11-13]

The standard bis-demethoxy curcumin, de-methoxy curcumin and curcumin was identified at R_f 0.29, 0.39 and 0.49 respectively.

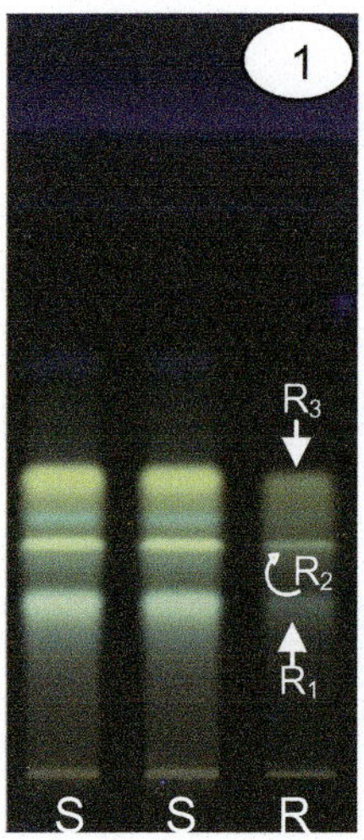

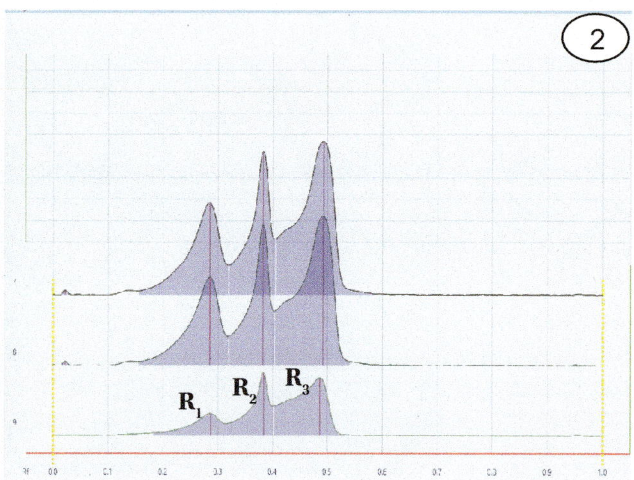

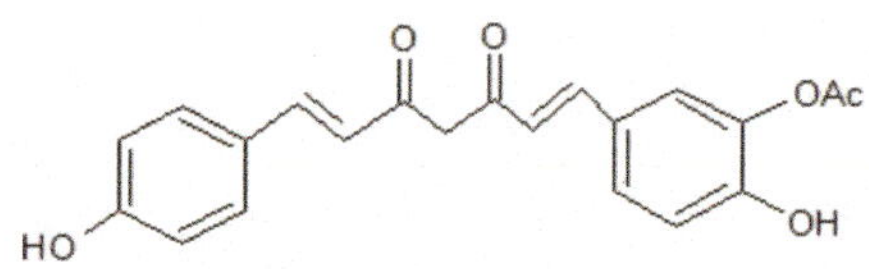

De-methoxy curcumin

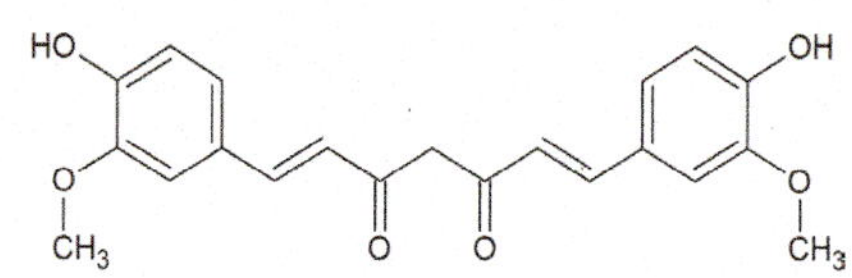

Curcumin

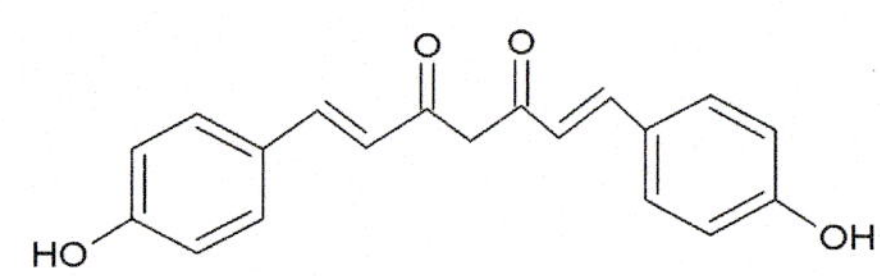

bis-demethoxy curcumin

Plate 1: HPTLC Chromatogram of *Curcuma aromatica* (S) and Standard Bis-demethoxy Curcumin (R1), Demethoxy Curcumin (R2), Curcumin (R3).

Plate 2: Overlay Spectra of Sample and Standard.

1. Ravindran PN, Nirmal BK, Sivaraman K (2007). Turmeric: The golden spice of life. In: Turmeric: The genus *Curcuma*. Boca Raton, FL, USA: CRC Press 11.
2. Kumar V, Sikarwar RLS (2002). Observations on some rare and endangered plants of Chattisgarh state, India. Phytotaxonomy 2: 135-142.
3. Srivastava S, Srivastava M, Rawat AKS, Mehrotra S (2006). Pharmacognostic evaluation of *Curcuma amada* Roxb. Proc Natl Acad Sci India 76: 153-160.
4. Anoop K (2015). Curcuma aromatica Salisb.: A Multifaceted Spice. Int J Phyto Res 6(1):10-15.
5. Quality standards of Indian medicinal plants (2008). Edn 1, Vol. 6, Indian Council of medical Research, Ramalingaswami Bhawan: 102- 109.
6. Choudhury SN, Anil CG, Madhumita S, Mina C, Piet AL (1996). Volatile Constituents of the Aerial and Underground Parts of *Curcuma aromatica* Salisb. from India. Journal of Essential Oil Research 8: 633-638.
7. Kumar A, Chomwal R, Kumar P, Renu S (2009). Antiinflammatory and wound healing activity of *Curcuma aromatica* salisb extract and its formulation. J Chem Pharma Res 1(1): 304-310.

8. Al-Reza SM, Rahman A, Sattar MA, Rahman MO, Fida HM (2010). Essential oil composition and antioxidant activities of *Curcuma aromatica* Salisb. Food Chem Toxicol 48: 1757.

9. Hu B, Shen KP, An HM, Wu Y, Du Q (2011). Aqueous extract of *Curcuma aromatica* induces apoptosis and G2/M arrest in human colon carcinoma LS-174-T cells independent of p53. Cancer Biother Radiopharm 26(1): 97-104.

10. Choochote W, Chaiyasit D, Kanjanapothi D, Rattanachanpichai E, Jitpakdi A, Tuetun B, Pitasawat B (2005). Chemical composition and anti-mosquito potential of rhizome extract and volatile oil derived from *Curcuma aromatica* against *Aedes aegypti* (Diptera: Culicidae). Journal of vector ecology 30(2): 302.

11. Sharad Srivastava, A. K. S. Rawat and S. Mehrotra (2007): Pharmacognostic evaluation of rhizome of *Curcuma aromatica* Salisb. Nigerian Journal of Nat Prod Med, 11: 10-14.

12. Sharad Kumar Srivastava, Ajay Kumar Singh Rawat, Shanta Mehrotra and P. Pushpangadan (2007): Botanical standardization of three commercially important *Curcuma* species. J Sci and Ind Res 66: 450-456.

13. Sharad Srivastava, Manjoosha Srivastava, Shanta Mehrotra and AKS Rawat (2009): A HPTLC Method for Chemotaxonomic evaluation of different *Curcuma* Species and their Commercial Samples. J Sci and Ind Res 68:876-880.

23

Curcuma longa Linn.

Family

Zingiberaceae

Vernacular name

Sanskrit: Ameshta; Hindi: Haldi; English: Indian saffron; Bengali: Halud; Telugu: Haridra.

Distribution [1]

The species is native plants to South Asia and is widely distributed in China and India. In India it is found in West Bengal, Tamil Nadu and Maharashtra.

Traditional Uses [2–3]

Traditionally *turmeric* is used against biliary disorders, anorexia, cough, diabetic wounds, hepatic disorders and rheumatism. The rhizome powder is used to treat diarrhea, roundworm, liver and stomach disease.

Major Chemical Constituents [4–5]

Curcumin, de-methoxy curcumin, bis-demethoxy curcumin, zingiberone, caffeic acid.

Pharmacological Activity [6–9]

Hepatoprotective, anti-microbial and anti-inflammatory activity.

Morphological Description

Curcuma longa linn. is commonly known as

A Vegetative Plant of *Curcuma longa*

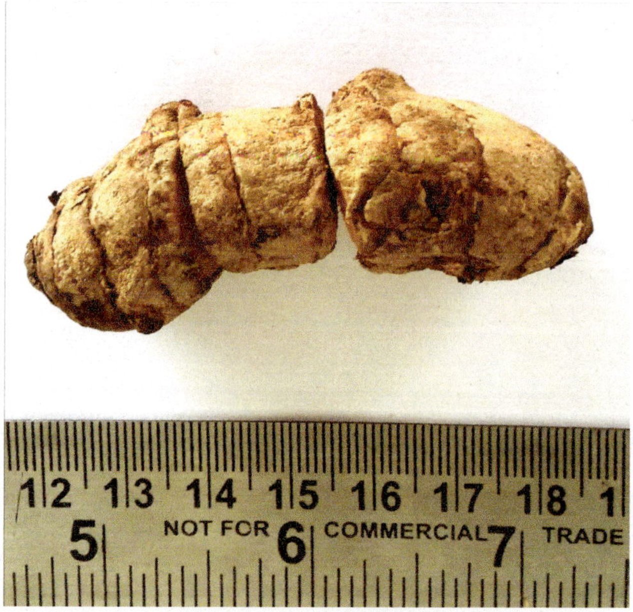

Dried Piece of Rhizome

turmeric. It is an herbaceous perennial plant, erect or semi-erect, rhizomatous, aromatic herb grows up to 1 meter. The leaves are long, petiolate, oblong-lanceolate shaped with pointed leaf apex, glaberous or hairy with prominent mid-vein and numerous lateral veins. Flowers are yellow they grouped together in dense spike. No fruiting is reported from this plant. The rhizome is large and branched, rough, segmented, fleshy and divided into nodes and internodes. At rhizome nodes scaly leaves are arranged circularly giving the appearance of growth rings with scars on the surface.

Anatomy

Transverse section of rhizome is almost circular in outline, show very narrow range of cork and wider cortex occupying major area of section, well distinguished endodermal layer and wide stellar system throughout the ground tissue. Detailed structure accounted that the rhizome shows multi-layered thin-walled cork cells made up of 4-6 layers. The cortex made with rounded or penta and hexacyclic parenchymatous cells contain scattered vascular bundle and oleoresin. Cortex and wide central ground tissue is parenchymatous and stealer vascular bundle is scattered throughout the ground tissue. Well defined single layer endodermis is present. Steles are collateral, non-lignified fibro-vascular bundles; they have been more numerous adjacent to periphery of endodermis. Stellar vascular bundles are well developed and composed of 2 to 10 vessels associated with narrow phloem. Oleoresin canals and simple oval to rod shaped starch grains are present throughout the ground tissue as well as in cortex region.

HPTLC Studies

Sample Solution

Accurately weight (5 g) crude powdered sample was subjected to cold extraction with methanol (50 ml), continuously shaken for 6 h and then allowed to stand at room temperature (25 °C ± 2) for 18 h. The extraction was repeated thrice, filtered and the pooled filtrate was concentrate in rotatory evaporator (Buchi, Switzerland). Working solution of strength 10 mg/ml was freshly prepared on the day of quantative analysis.

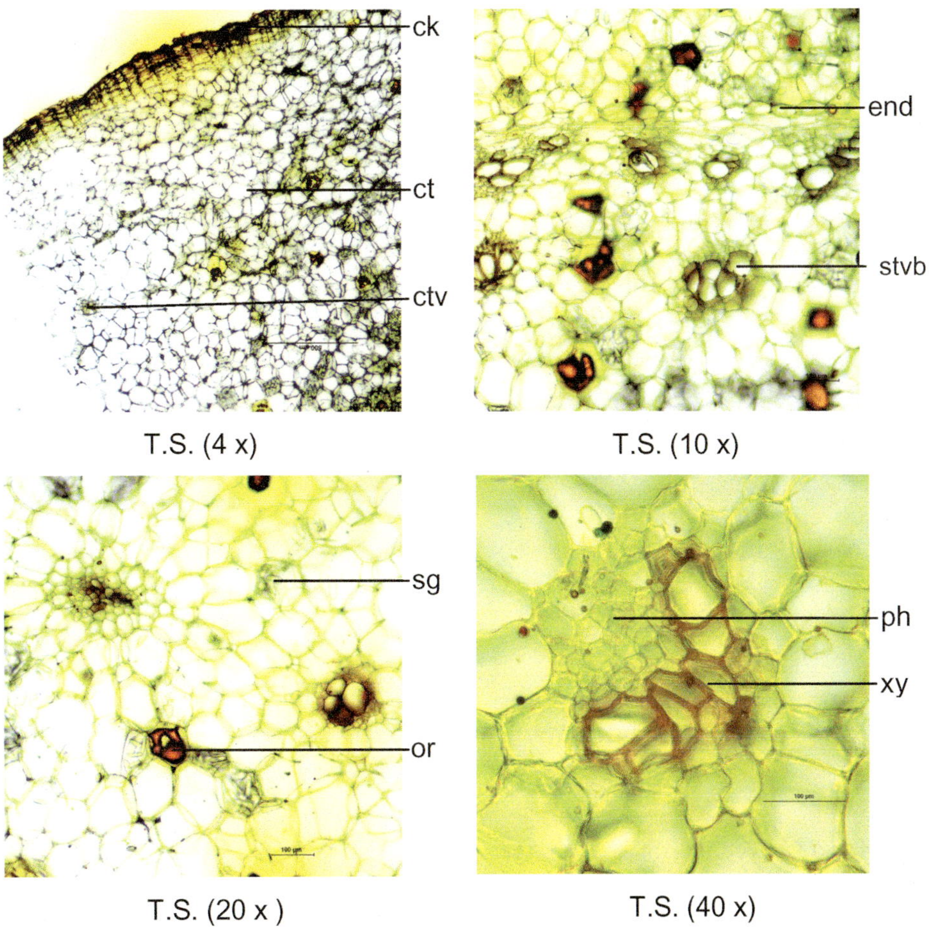

Transverse Section (T.S.) of *Curcuma longa* Rhizome.

Abbreviation: ck- cork; ct- cortex; ctvb- cortical vascular bundle; end- endodermis; orc- oleoresin canal; ph- phloem; sg- starch grain; stvb- stellar vascular bundle; xy- xylem.

Standard Solution

Curcumin (≥ 98 per cent), de-methoxy curcumin (≥ 98 per cent) and bis-demethoxy curcumin (≥ 98 per cent) was freshly prepared of strength 1 mg/ml.

Sample Application

2 µl of standard solution and 10 µl of sample was applied as 6 mm band with CAMAG HPTLC applicator on activated silica gel 60 GF_{254} HPTLC plate. The chromatogram was developed on pre-saturated CAMAG HPTLC twin trough chamber for 15 min and the solvent system was allowed to run for 8 cm.

Solvent System

Chloroform- ethanol- glacial acetic acid (9.4:0.5:0.1 v/v)

Visualization

The chromatogram was visualized at UV_{366} and densitometric scanning of curcumin, de-methoxy curcumin and bis-demethoxy curcumin was done at 420 nm.

Identification of Reference Markers

The standard bis-demethoxy curcumin, de-methoxy curcumin and curcumin was identified at R_f 0.29, 0.39 and 0.49 respectively.

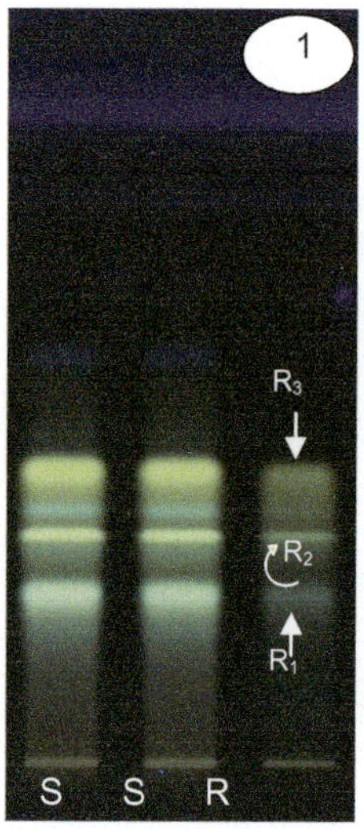

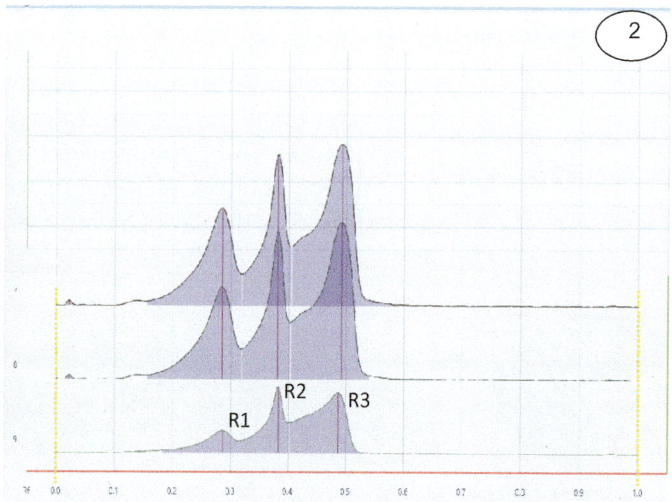

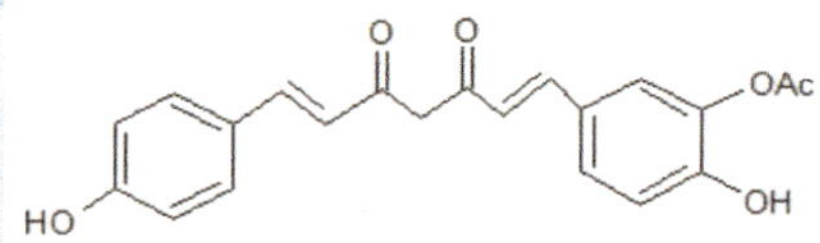

De-methoxy curcumin

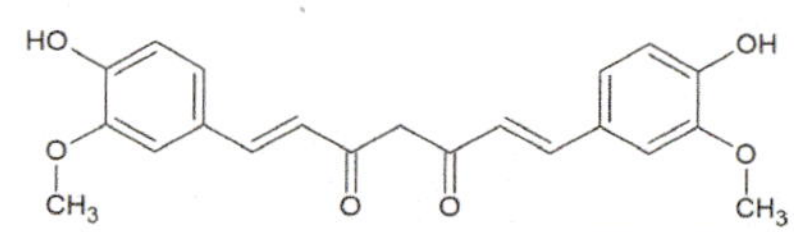

Curcumin

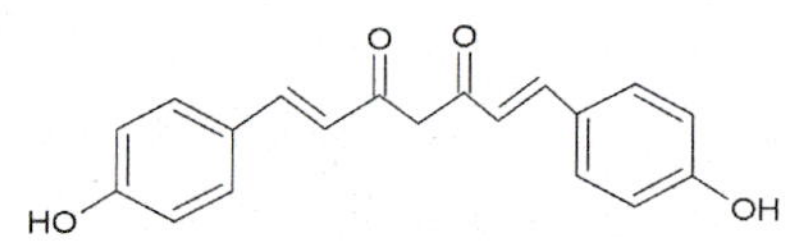

Bis-demethoxy curcumin

Plate 1: HPTLC Chromatogram of *Curcuma longa* (S) and Standard Bis-demethoxy Curcumin (R1), Demethoxy Curcumin (R2), Curcumin (R3).

Plate 2: Overlay Spectra of Sample and Standard.

1. Kurup PNV (1977). Handbook of Medicinal Plants, Vol. I, Central Council for Research in Indian Medicine and Homoeophathy (CCRIMH), New Delhi.

2. Nasri H, Sahinfard N, Rafieian M, Rafieian S, Shirzad M, Rafieian-kopaei M (2014). Turmeric, A spice with multifunctional medicinal properties. J Herbmed Plarmacol 3(1): 5-8.

3. Yadav PR, Tarun G (2017). Versatility of turmeric: A review the golden spice of life. J Pharma Phyto 6(1): 41-46.

4. Sharma RA, Gescher AJ, Steward WP (1955). Curcumin, the story so far. Eur J Cancer 41: 1955-68.

5. Pfeiffer E, Hhle S, Solyom AS, Metzler M (2003). Studies on the stability of turmeric constituents. J Food Enginee 56: 257-259.

6. Ruby J, Kuttan G, Babu KD, Rajashekharan KN, Kuttan R (1995). Antitumor and oxidant activity of natural curcuminoids. Cancer Lett 94: 79-83.

7. Kim GY, Kim KH, Lee SH, Yoon MS, Lee HJ, Moon DO (2005). Curcumin inhibits immunostimulatory function of dendritic cells, MAPKs andtranslocation of NF-B as potential targets. J Immunol 174: 8116-24.

8. Chattopadhyay I, Kaushik B, Uday B, Ranajit KB (2004). Turmeric and curcumin, Biological actions and medicinal applications. Current Science 87(1): 44-53.

9. Arutselvi R, Balasaravanan T, Ponmurugan P, Muthu Saranji N, Suresh P (2012). Phytochemical screening and comparative study of anti microbial activity of leaves and rhizomes of turmeric varieties. Asian J Plant Sci Res 2(2): 212-219.

24

Curcuma zedoaria Rosc.

Family

Zingiberaceae

Vernacular name [1]

Sanskrit: Krachura; Hindi: Gandamatsi; Bengali: Sutha; English: White turmeric, yellow zedoary; Chinese: Er-jyur.

Distribution [1]

Curcuma zedoaria is indigenous to India, Srilanka, Bangladesh and widely cultivated in Nepal, Thailand and China. In India it is distributed in wild throughout north-east, central and south India.

Traditional Uses [1-2]

The rhizome of *Curcuma zedoaria* is used in folk medicine to treat cough and cold, vomiting, wounds, dysentery, ûatulence, skin diseases and worm infection in child.

Major Chemical Constituents [3].

Curcumin, de-methoxy curcumin, bis-demethoxy curcumin

Pharmacological Activity [4-7]

Anticancer, antifungal, antibacterial and anti-inflammatory activity

A Flowering Stalk with Rhizome

Dried Piece of Rhizome

Morphological Description

Curcuma zedoaria Rosc. is commonly known as white turmeric and yellow zedoary. It is an herbaceous perennial plant, erect or semi-erect, rhizomatous, aromatic herb. The leaves are long, petiolate, oblong–lanceolate shaped with pointed leaf apex, glaberous or hairy with prominent mid-vein and numerous lateral veins. Flowers are pinkish and grouped together in dense spike. The rhizome is conical or top shaped, large and branched, rough, segmented, fleshy and attached to the primary rhizome are several sessile finger shaped lateral branches. At rhizome nodes scaly leaves are arranged circularly giving the appearance of growth rings with scars on the surface.

Anatomy

Transverse section of rhizome is almost circular in outline, show very narrow range of cork and wider cortex occupying major area of section, well distinguished endodermal layer and wide stellar system throughout the ground tissue. Detailed structure accounted that the rhizome shows multi-layered thin-walled cork composed of 7-10 rows of rectangular to tangentially elongate cells. The ground tissue is differentiated into two regions, the outer cortex and the inner cortex by a distinct endodermis. The cortex made with rounded or penta and hexacyclic parenchymatous cells contain scattered vascular bundle and oleoresin. Cortex and wide central ground tissue is parenchymatous and stealer vascular bundle is scattered throughout the ground tissue. Steles are collateral, non-lignified fibro-vascular bundles; they have been more numerous adjacent to periphery of endodermis. Stellar vascular bundles are well developed and composed of 2 to 5 vessels associated with narrow phloem. Oleoresin canals and simple oval to rod shaped starch grains are present throughout the ground tissue as well as in cortex region.

HPTLC Studies

Sample Solution

Accurately weight (5 g) crude powdered sample was subjected to cold extraction with methanol (50 ml), continuously shaken for 6 h and then allowed to stand at room temperature (25 °C ± 2) for 18 h. The extraction

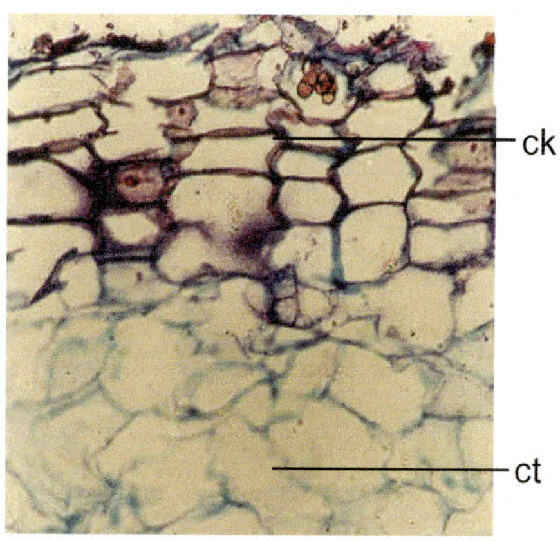

T.S. (40 x)

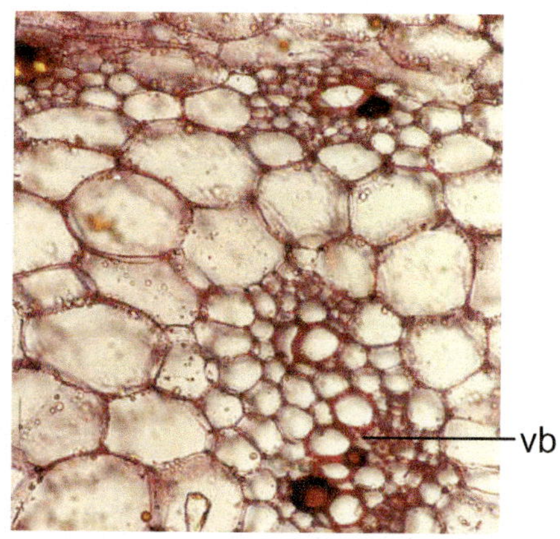

T.S. (20 x)

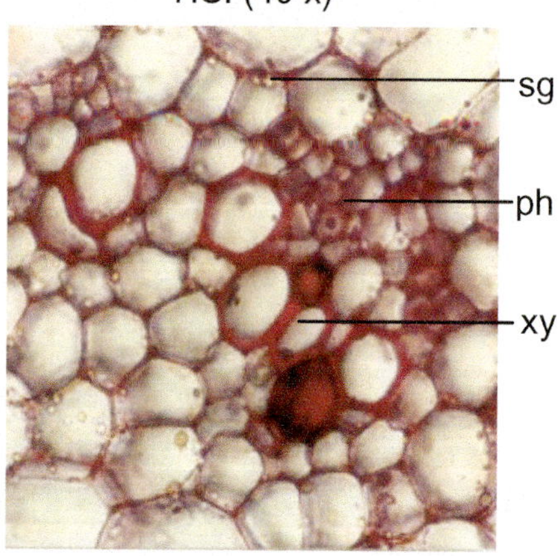

T.S. (40 x)

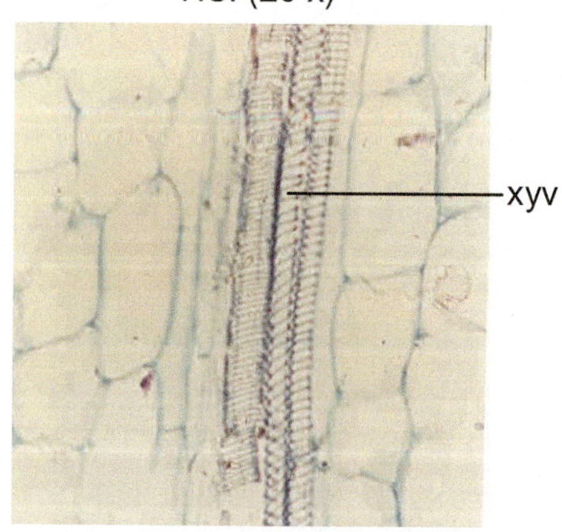

L.S. (40 x)

Transverse Section (T.S.) & Longitudinal Section (L.S.) of *Curcuma zedoaria* Rhizome.

Abbreviation: ck- cork; ct- cortex; ph- phloem; sg- starch grains; vb- vascular bundle; xy- xylem; xyv- xylem vessels.

was repeated thrice, filtered and the pooled filtrate was concentrate in rotatory evaporator (Buchi, Switzerland). Working solution of strength 10 mg/ml was freshly prepared on the day of quantative analysis.

Standard Solution

Curcumin (≥ 98 per cent), de-methoxy curcumin (≥ 95 per cent), bis-demethoxy curcumin (≥ 95 per cent) was freshly prepared of strength 1 mg/ml.

Sample Application

2 μl of standard solution and 10 μl of sample was applied as 6 mm band with CAMAG HPTLC applicator on activated silica gel 60 GF_{254} HPTLC plate. The chromatogram was developed on pre-saturated Camag HPTLC twin trough chamber for 15 min and the solvent system was allowed to run for 8 cm.

Solvent System

Chloroform: ethanol: glacial acetic acid (9.4:0.5:0.1 v/v)

Visualization

The chromatogram was visualized at UV_{366} and densitometric scanning of Curcumin, de-methoxy curcumin and bis-demethoxy curcumin was done at 420 nm.

Identification of Reference Markers[9-11]

The standard bis-demethoxy curcumin, de-methoxy curcumin and curcumin was identified at R_f 0.29, 0.39 and 0.49 respectively.

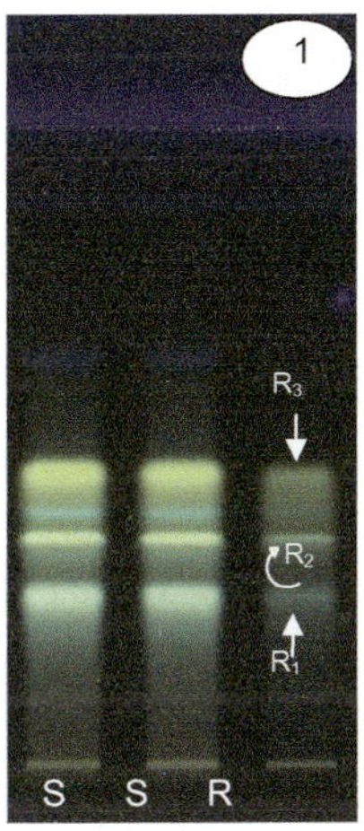

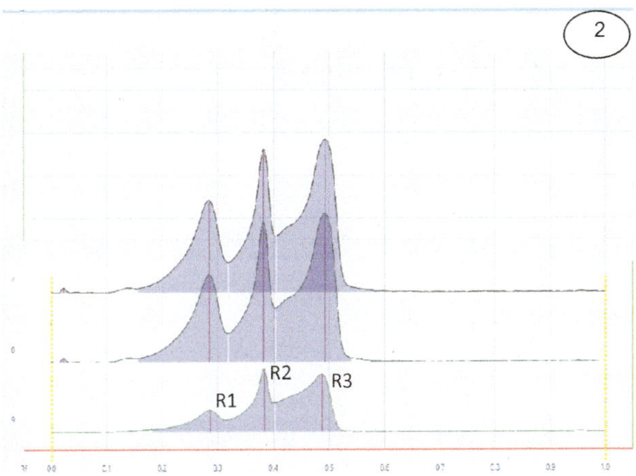

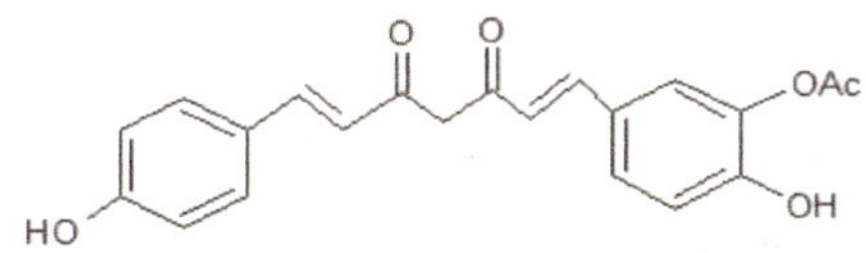

De-methoxy curcumin

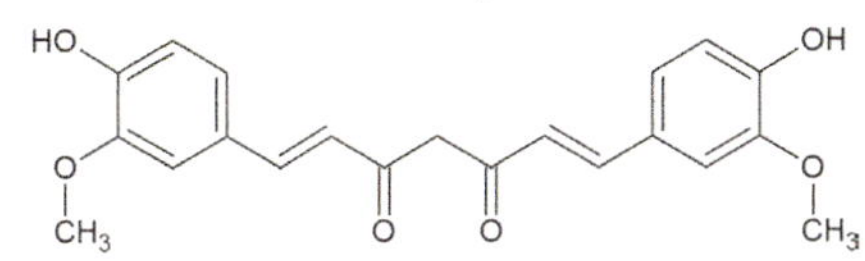

Curcumin

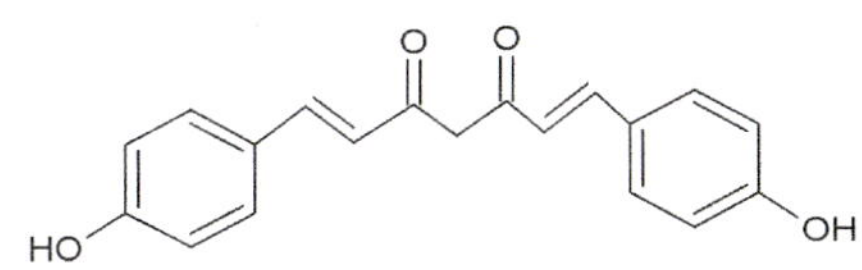

bis-demethoxy curcumin

Plate 1: HPTLC Chromatogram of *Curcuma zedoaria* (s) and Standard bis-demethoxy curcumin (R1), Demethoxy Curcumin (R2) and Curcumin (R3).

Plate 2: Overlay Spectra of Sample and Standards.

1. The standard bis-demethoxy curcumin, de-methoxy curcumin and curcumin was identified at R_f 0.29, 0.39 and 0.49 respectively.
2. Nadkarni KM (1999). Indian Materia Medica, 3rd edn. Mumbai, India: Popular Prakashan Private Limited.
3. Pitasawat B, Champakaew D, Choochote W, Jitpakdi A, Chaithong U, Kanjanapothi D, Rattanachanpichai E, Tippawangkosol P, Riyong D, Tuetun B (2007). Aromatic plant-derived essential oil: An alternative larvicide for mosquito control. Fitoterapia 78: 205-210.
4. Syu WJ, Shen CC, Don MJ, Ou JC, Lee GH, Sun CM (1998). Cytotoxicity of curcuminoids and some novel compounds from *Curcuma zedoaria*. J Nat Pro 61(12): 1531-4.
5. Seo WG, Hwang JC, Kang SK, Jin UH, Suh SJ, Moon SK, Kim CH (2005). Suppressive effect of Zedoariae rhizome on pulmonary metastasis of B16 melanoma cells. J. Ethnopharmacol 101: 249–257.
6. Makabe H, Maru N, Kuwabara A, Kamo T, Hirota M (2006). Anti-inûammatory sesquiterpenes from *Curcuma zedoaria*. Nat Prod Res 20: 680-685.
7. Ficker CE, Smith ML, Susiarti S, Leaman DJ, Irawati C, Arnason JT (2003). Inhibition of human pathogenic fungi by members of Zingiberaceae used by the Kenyah (Indonesian Borneo). J ethnophar 85(2-3):289-93.

8. Bugno A, Nicoletti MA, Almodóvar AA, Pereira TC, Auricchio MT (2007). Antimicrobial efficacy of *Curcuma zedoaria* extract as assessed by linear regression compared with commercial mouthrinses. Brazilian J 38(3): 440-5.

9. Sharad Srivastava, A. K. S. Rawat and S. Mehrotra (2011). Pharmacognostic evaluation of the rhizomes of *Curcuma zedoaria* Rosc. Pharmacog J 3(20): 20-26.

10. Sharad Srivastava, Manjoosha Srivastava, Shanta Mehrotra and AKS Rawat (2009). A HPTLC Method for Chemotaxonomic evaluation of different *Curcuma* Species and their Commercial Samples. JJ Sci Ind Res, 68:876-880.

11. Sharad Kumar Srivastava, Ajay Kumar Singh Rawat, Shanta Mehrotra and P. Pushpangadan (2007). Botanical standardization of three commercially important *Curcuma* species. J Sci Ind Res 66: 450-456.

25

Cyperus rotundus Linn.

Family

Cyperaceae

Vernacular Name [1]

Sanskrit: Muthakasu, Musta; Hindi: Motha, Nagagmotha; English: Nut grass, Java-grass, Purple nutsedge; Bengali: Nagarmotha; Malayalam: Korakizanna; Marathi: Barik motha.

Distribution [2]

It is a weed found in tropical, subtropical and temperate regions of the world. *C. rotundus* is native to Africa, southern Asia; it is distributed throughout India as a weed.

Traditional Uses [3-5]

C. rotundes is traditionally used as carminative, astringent, diuretic, hepatoprotective, antipyretic. It is also used for gastrointestinal spasms, stomach disorders, vomiting, intestinal parasites and food poisoning. In Ayurveda, its tuber was considered as astringent, diaphoretic, diuretic, analgesic, antispasmodic, aromatic and carminative.

A Flowering Plant

Dried Tuber

Major Chemical Constituents [6-8]

Oleanolic acid, rotundone, linoleic acid, salicylic acid and caffeic acid.

Pharmacological Activity [9-11]

Anti-dirrohoeal, antimicrobial, antioxidant and anti-inflammatory activity.

Morphological Description

C. rotundus is a perennial, erect, slender herb which spread by fibrous root system. Stem are smooth, erect, triangular in shape, thickened nodosely at the base and constricted into wiry rhizome. Leaves are arises from the base of the plant; they are shiny, dark green, long and narrow, often overlapping stem, smooth upper surface and pointed towards apex. Flowers are produced in a cluster at the end of stem with reddish-brown to purple colour. Tuber are small, light brownish, texture hard, rounded to oval, arises from the rhizome, which stores food reserve in the form of starch. Seed are up to 2 mm long and black to brown colour.

Anatomy

Transverse section of rhizome tuber is almost circular in outline, show very narrow range of cork and wider cortex occupying major area of section, well distinguished endodermal layer and wide stellar system throughout the ground tissue. Detailed structure accounted that the rhizome shows multi-layered thin-walled cork cells made up of 4-6 or more layers. The cortex made with rounded or pentacyclic parenchymatous cells contain starch grains. Cortex and wide central ground tissue is parenchymatous and stealer vascular bundle is scattered throughout the ground tissue. Well defined single layer endodermis is present. Steles are collateral, lignified fibro-vascular bundles; they have been more numerous adjacent to periphery of endodermis. Stellar vascular bundles are well developed and present in a ring

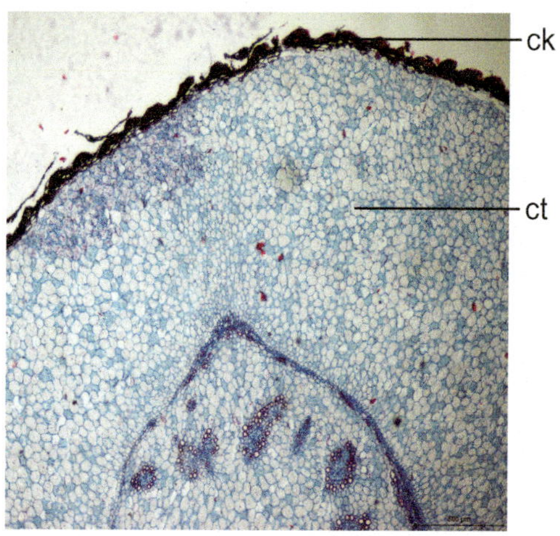

T.S. (4 x)

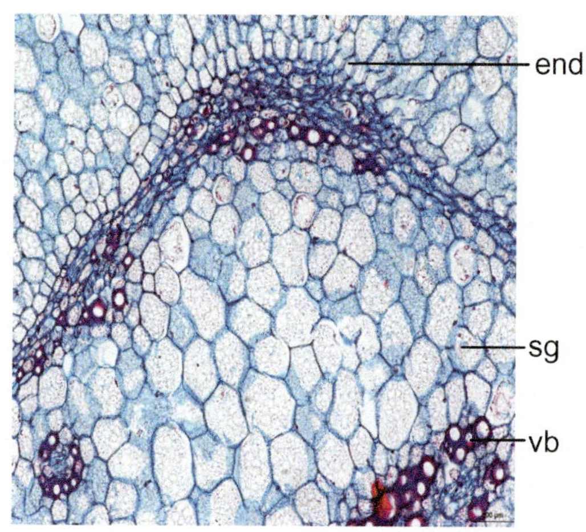

T.S. (10 x)

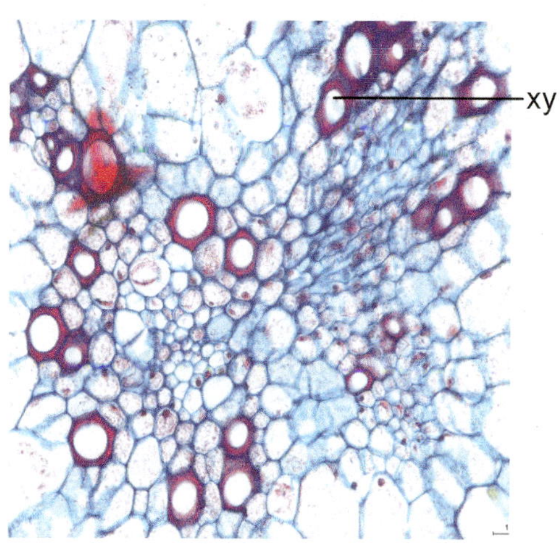

T.S. (20 x)

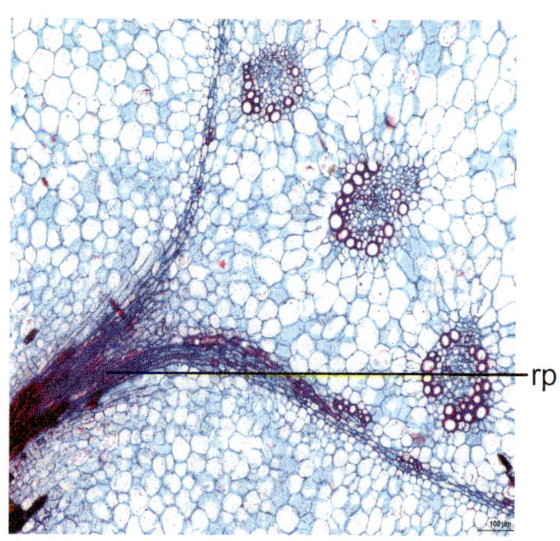

T.S. (10 x)

Transverse Section (T.S.) of *Cyperus rotundus* Tuber.

Abbreviation: ck- cork; ct- cortex; end- endodermis; rp- root primordium; sg- starch grains; vb- vascular bundle; xy- xylem.

form, composed of 6 to 20 or more vessels associated with narrow phloem. Simple rounded to oval shaped starch grains are present throughout the ground tissue as well as in the cortical region.

HPTLC Studies

Working Solution

Accurately weight (5 g) crude powdered sample was extracted with methanol (50 ml) followed by continuously shaken for 6 h and then allowed to stand for 18 h at room temperature (25 ± 2^{0}C). Extraction was repeated thrice and the pooled filtrate was concentrate in rotatory evaporator (Buchi, Switzerland), dried residue was stored at 4 °C. Working solution of 10 mg/ml was freshly prepared on the day of quantitative analysis.

Standard Solution

Oleanolic acid (≥ 98 per cent) was freshly prepared of strength 1 mg/ml.

Sample Application

2 µl of standard solution and 10 µl of sample was applied as 6 mm band with CAMAG HPTLC applicator on activated silica gel 60 GF$_{254}$ HPTLC plate. The chromatogram was developed on pre-saturated Camag HPTLC twin trough chamber for 20 min and the solvent system was allowed to run for 8 cm.

Solvent System

Toluene- ethyl acetate- formic acid (7:3:0.5 v/v).

Visualization

The chromatogram was visualized after derivatization with anisaldehyde sulphuric acid reagent and densitometric scanning was done at 530 nm.

Identification of Oleanolic Acid[12]

The standard oleanolic acid was identified at R_f 0.67.

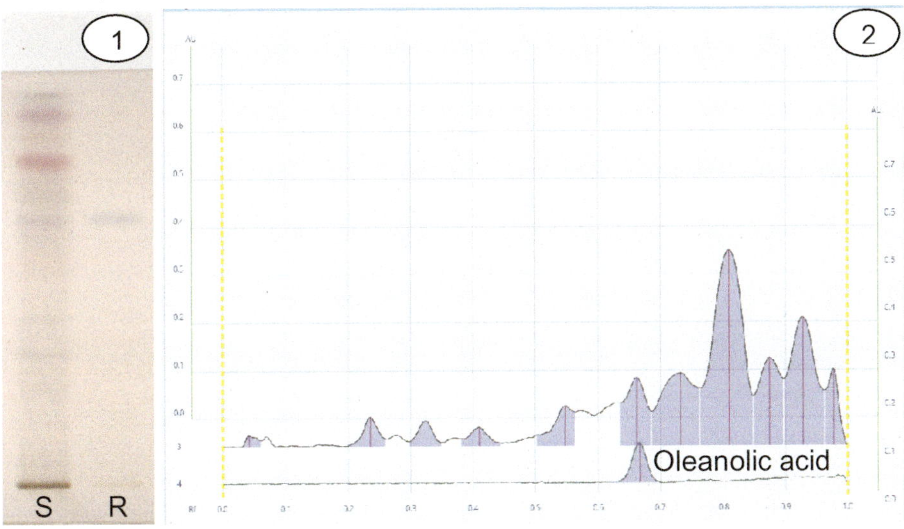

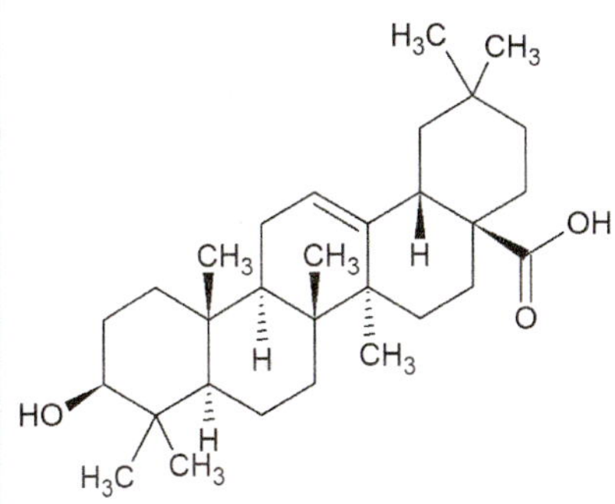

Oleanolic acid

Plate 1: HPTLC Chromatogram of *Cyperus rotundus* (S) and Standard Oleanolic Acid (R), Visualized After Derivatization with Anisaldehyde Sulphuric Acid Reagent.

Plate 2: Overlay Spectra of Sample and Standard.

1. Kirtikar KR, Basu BD (1996). Indian Medicinal Plants, Valley Offset Printers and Publishers Dehradun 4:2638-2640.
2. Parotta JA (2001). Healing Plants of Peninsular India. New York: CABI Publishing 02 66.
3. Srivastava S, Mishra G (2015). *Cyperus rotundus* (Nagarmotha)-An Important Medicinal Plant: A Review. IJPR 1(1):11-22.
4. Chopra RN, Nayar SL, Chopra IC (1986). Glossary of Indian medicinal plants. Council of Scientific and Industrial Research, New Delhi.
5. Sivapalan SR (2013). Medicinal uses and pharmacological activities of *Cyperus rotundus* Linn - a review. International Journal of Scientific and Research Publications 3(5): 1-8.
6. Bisht A, Bisht GR, Singh M, Gupta R, Singh V (2011). Chemical composition and antimicrobial activity of essential oil of tubers of *Cyperus rotundus* Linn. collected from Dehradun (Uttarakhand). Int J Res Pharm Biomed Sci 2:661 5.
7. Singh A, Singh N (2016). Ethno-Phamaco-therapeutic Activities of *Cyperus rotundus*. IJMAS 3(2):186-194.

8. Huang KC (1999). The dried tuber of *Cyperus rotundus* L. The pharmacology of Chinese herbs, 2nd ed.: 320-321.

9. Daswani PG, Brijesh S, Tetali P, Birdi TJ (2011). Studies on the activity of *Cyperus rotundus* Linn. tubers against infectious diarrhea. Indian J Pharmacol 43: 340-4.

10. Nagulendran kr, Velavan S, Mahesh R (2007). In Vitro Antioxidant Activity and Total Polyphenolic Content of *Cyperus rotundus* Rhizomes. E-Journal of Chemistry 4(3): 440-449.

11. Seo WG, Pae HO, Oh GS, Chai KY, Kwon TO, Yun YG (2001). Inhibitory effects of methanol extract of *Cyperus rotundus* Linn. rhizomes on nitric oxide and superoxide productions by murine acrophage cell line, RAW 264.7 cells. J Ethnopharmacol 76(1): 59-64.

12. Priyanka Dhar, Debasmita Ghosh Dhar, AKS Rawat, Sharad Srivastava (2017). Medicinal chemistry and biological potential of *Cyperus rotundus* Linn.: An overview to discover elite chemotype(s) for industrial use. Ind Crop Prod 108: 232-247.

26

Desmodium gangeticum (L.) DC.

Family

Fabaceae

Vernacular Name

Sanskrit: shalaparni, anshumati; Hindi: salpani; Bengali: salpani; Marathi: radbhal; Telugu: gitanaram, kolakuporna; Gujarathi: selman.

Distribution [1]

Desmodium gangeticum is spread to Sri Lanka, China, Philippine and tropical Africa. In India it is distributed throughout the plain areas and lower hills up to 5000 ft.

Traditional Uses [2-3]

It is used in folk medicine to treat fever, inhibit pain, cough, diuretic, anthelmintic, laxative. Pawara tribals of Satpuda Hills, Maharashtra, India root powder and honey are applied frequently to treat mouth ulcer.

Major Chemical Constituents [4-5]

Lupeol, stigmasterol, gangetin, desmodin and rutin.

Pharmacological Activity [5-8]

Anti-diabetic, antioxidant, anti-inflammatory, anti-leishmanial, immunomodulatory and anti-nociceptive activity

A Vegetative Plant

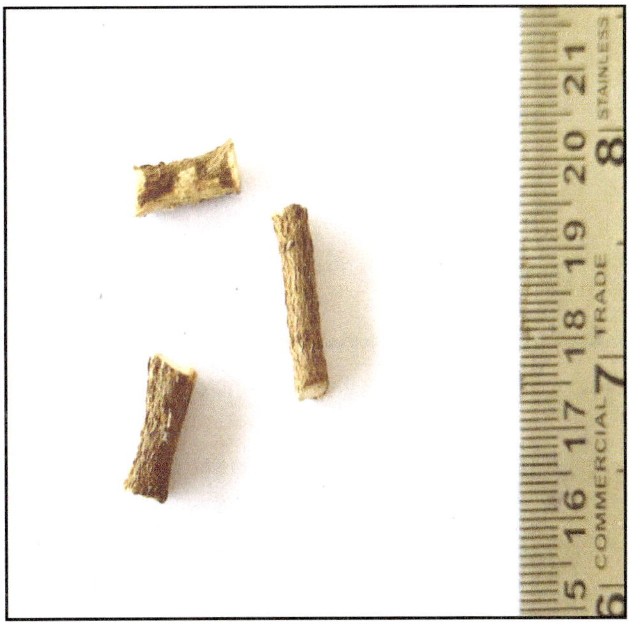

Dried Pieces of Root

Morphological Description

Desmodium gangeticum is a perennial erect shrub growing up to 2 to 4 feet. Stem is woody, erect, and cylindrical with numerous irregular branches. Leaves are small, alternate, unifoliate, petiolate, ovate-oblong or rounded, glabrous covered with dense soft hairs. Inflorescence is terminal, flowers are small, purple or white colored and seed are kidney shaped. The root is poorly developed; smooth, light yellow in colour and bark is yellowish white and has a leathery texture and bitter taste.

Anatomy

Transverse section of root is circular in outline, show very narrow range of cork, wider cortex and well developed wide stellar system. Detailed structure accounted that the root shows narrow cork, followed by multi-layered thin-walled cork cambium lie underneath this. The cortical region is made up of 15 to 18 layered irregular parenchymatous cells, containing round or oval starch grains. Just below the cortex phloem are present having sieve tubes, companion cells and phloem fibers. The xylem is 8-10 celled wide and consists of vessels, fibers, tracheids and xylem parenchyma. Vessels are mostly in groups of 5 to 8 and rarely solitary. The medullary rays are thick-walled, multiseriate, radiating, and parenchymatous containing starch grains.

HPTLC Studies

Working Solution

Accurately weigh (5 g) crude powdered sample was extracted (cold) with methanol (50 ml) followed by continuously shaken for 6 h and then allowed to stand for 18 h at room temperature (25 ± 2°C). The extraction was repeated thrice and the pooled filtrate was concentrate in rotatory evaporator (Buchi, Switzerland), dried residue was stored at 4 °C. Working solution of 10 mg/ml was freshly prepared on the day of quantative analysis.

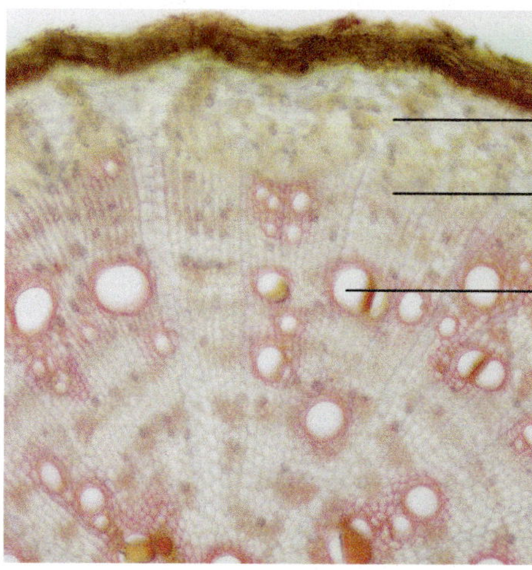

T.S. (10 x)

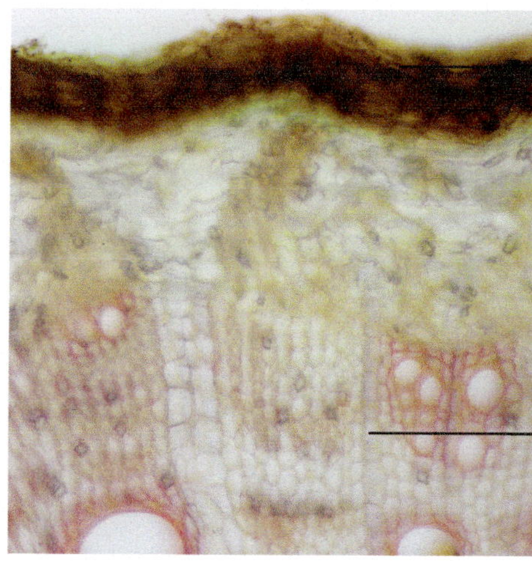

T.S. (20 x)

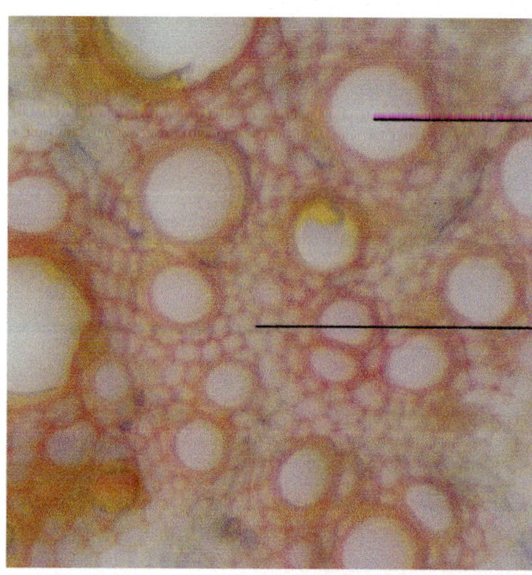

T.S. (40 x)

Transverse Section (T.S.) of *Desmodium gangeticum* Root.

Abbreviation: ck- cork; ct- cortex; cjt- conjunctive tissues; mr- medullary rays; ph- phloem; xy- xylem.

Standard Solution

Lupeol (≥ 98 per cent) was freshly prepared of strength 1 mg/ml.

Sample Application

2 µl of standard solution and 10 µl of sample was applied as 6 mm band with CAMAG HPTLC applicator on activated silica gel 60 GF_{254} HPTLC plate. The chromatogram was developed on pre-saturated CAMAG HPTLC twin trough chamber for 10 min and the solvent system was allowed to run for 8 cm.

Solvent System

Toluene- ethyl acetate- formic acid (7:3:0.5 v/v).

Visualization

The chromatogram was visualized after derivatization with anisaldehyde sulphuric acid reagent and densitometric scanning was done at 530 nm.

Identification of Lupeol

The standard lupeol was identified at R_f 0.82.

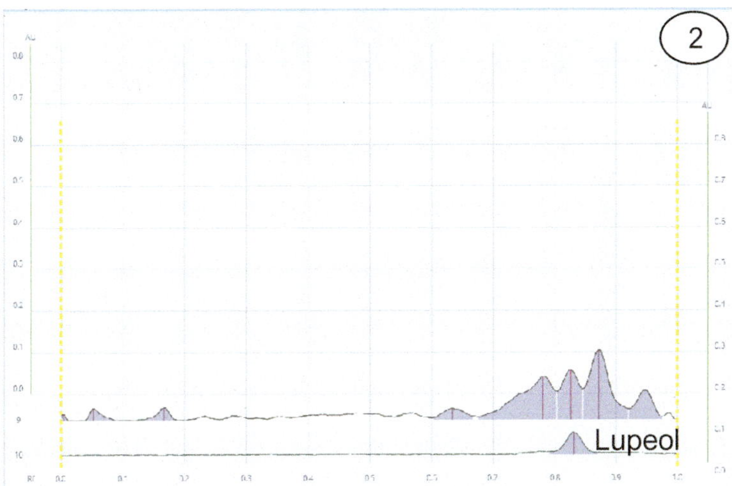

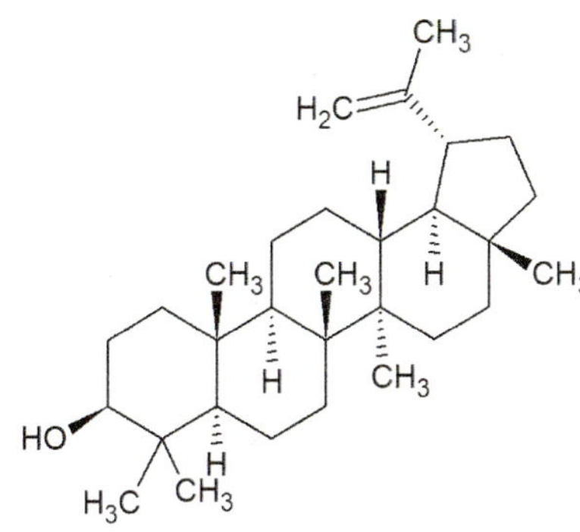

Lupeol

Plate 1: HPTLC Chromatogram of *Desmodium gangeticum* (S) and Standard Lupeol (R), Visualized After Derivatization with Anisaldehyde Sulphuric Acid Reagent.

Plate 2: Overlay Spectra of Sample and Standard.

1. Trout K (2002). Trout's Notes on the Genus Desmodium (Chemistry, Ethnomedicine, Pharmacology, Synonyms and Miscellany). Mydriatic Productions 2- 38.
2. Xueqin M, Chengjian Z, Changling H, Khalid R, Luping Q (2011). The genus *Desmodium* (Fabaceae)-traditional uses in Chinese medicine, phytochemistry and pharmacology. J Ethnopharmacol 138: 314-32.
3. Kosalge SB, Fursule RA (2009). Investigation of ethnomedicinal claims of some plants used by tribals of Satpuda Hills in India. J Ethnopharmacol 121: 456–61.
4. Subha R, Madan MP, Ajay KSR (2011). An ethnomedicinal, phytochemical and pharmacological profile of *Desmodium gangeticum* (L.) DC. and *Desmodium adscendens* (Sw.) DC. J Ethnopharmacol 136: 283-96.
5. Mishra PK, Singh N, Ahmad G, Dube A, Maurya R (2005). Glycolipids and other constituents from *Desmodium gangeticum* with antileishmanial and immunomodulatory activities. Bioorg Med Chem Lett 15: 4543-46.
6. Bisht R, Bhattacharya S (2013). Effect of various extracts of *Desmodium gangeticum* on Streptozotocin-nicotinamide induced type-2 diabetes. Asian J Plant Sci Res 3(3):28-34.
7. Kurian GA, Yagnesh N, Kishan RS, Paddikkala J (2008). Methanol extract of *Desmodium gangeticum* roots preserves mitochondrial respiratory enzymes, protecting rat heart against oxidative stress induced by reperfusion injury. J Pharm Pharmacol 60(4): 523-30.
8. Rathi A, Rao CV, Ravishankar B, Deb S, Mehrotra S (2004). Anti-inflammatory and anti-nociceptive activity of the water decoction *Desmodium gangeticum*. J Ethnopharmacol 95: 259-63.

27

Dioscorea deltoidea Wall. Ex Griseb.

Family

Dioscoreaceae

Vernacular name [1-2]

Sanskrit: Varahikand; Hindi: Gun, Kin; English: Wild yam, Elephant's foot; Punjabi: Kitra; Urdu: Qanis.

Distribution [1-2]

Dioscorea deltoidea is vulnerable in Red Data Book of Indian Plants, earlier It was assessed as critically endangered in 1998. In Asia the plant is found in India, China, Pakistan and Nepal. The species is found at an altitude of 2000-9000 feet in Indian Himalayas.

Traditional Uses [3-5]

Dioscorea deltoidea is used as haemostatic, diuretic, expectorant and expelling worm from the body. In Kumaun region of Uttarakhand rhizome powder is orally taken to cure dysentery, piles and abdominal pain.

Plant in Natural Habitat

Dried Powder

Major Chemical Constituents [6]

Diosgenin, β-Sitosterol, kaempferol

Pharmacological Activity [7-10]

Anti-tumor, anti-fungal, cytotoxic and anti-oxidant activity

Morphological Description

It is a tropical or subtropical herbaceous, perennial climber, growing up to 8 feet. Leaves are simple, broad, triangular ovate, heart shaped, long pointed, petiolate with prominent veins. Stem are cylindrical, herbaceous, smooth and green. Inflorescence is spike, solitary; male flower are cylindrical, solitary in leaves axils. Rhizomes are ligneous irregular, fleshy, tuberous and secondary roots arise from it.

Anatomy

Transverse section of rhizome is irregular in outline, show very narrow range of cork and wider cortex, occupying major area of section, well developed wide stellar system in ground tissue. Detailed structure accounted that the rhizome shows narrow cork made of multiple layer of irregular shape cell, followed by multi-layered thin-walled cork cells lie underneath this; cortex and wide central ground tissue is parenchymatous and stealer vascular bundle is scattered throughout in ground tissues. In cortex sclereids are oftenly seen, in solitary as well as in group. Steles are collateral, non-lignified fibro-vascular bundles; vascular bundles are well developed and composed of 1 to 5 vessels associated with narrow phloem. Simple oval to pear shaped starch grains are present throughout the ground tissue.

HPTLC Studies

Sample Solution

Accurately weighed (5 g) crude powdered rhizome was hydrolyzed with (50 ml) of 2.5 N hydrochloric acid for 4 hrs at 80 oC. The residue was filtered, washed with hot water until neutral or free from acid. The residue was then dried in hot air oven and further extracted with methanol (50 ml) by continuously shaken for 6 h and then allowed to stand for 18 h at room temperature (25 oC ± 2). Extraction was repeated thrice, filtered and the pooled filtrate was concentrated in rotatory evaporator (Buchi, Switzerland), dried residue was stored at 4 OC. Working solution of 10 mg/ml was freshly prepared on the day of analysis.

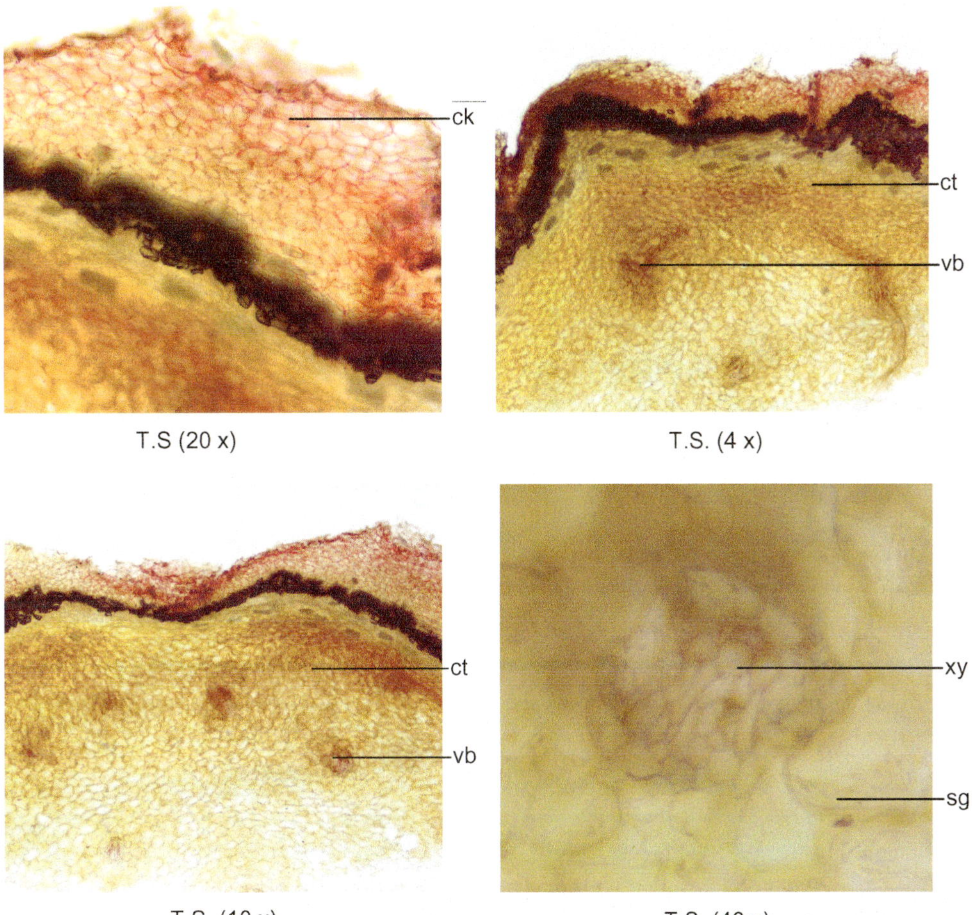

T.S (20 x) T.S. (4 x) T.S. (10 x) T.S. (40 x)

Transverse Section (T.S.) of *Diascorea deltoida* Rhizome.

Abbreviation: ck- cork; ct- cortex; sg- starch grains; vb- vascular bundle; xy- xylem.

Standard Solution

Diosgenin (≥ 98 per cent) was freshly prepared of strength 1 mg/ml.

Sample Application

2 µl of standard solution and 10 µl of sample was applied as 6 mm band with CAMAG HPTLC applicator on activated silica gel 60 GF_{254} HPTLC plate. The chromatogram was developed on pre-saturated CAMAG HPTLC twin trough chamber for 20 min and the solvent system was allowed to run for 8 cm.

Solvent System

Toluene- ethyl acetate- diethyl amine (6:2:0.3 v/v)

Visualization

The chromatogram was visualized after derivatization with anisaldehyde sulphuric acid reagent and densitometric scanning of solasodine was done at 440 nm.

Identification of Diosgenin

The standard diosgenin was identified at R_f 0.45.

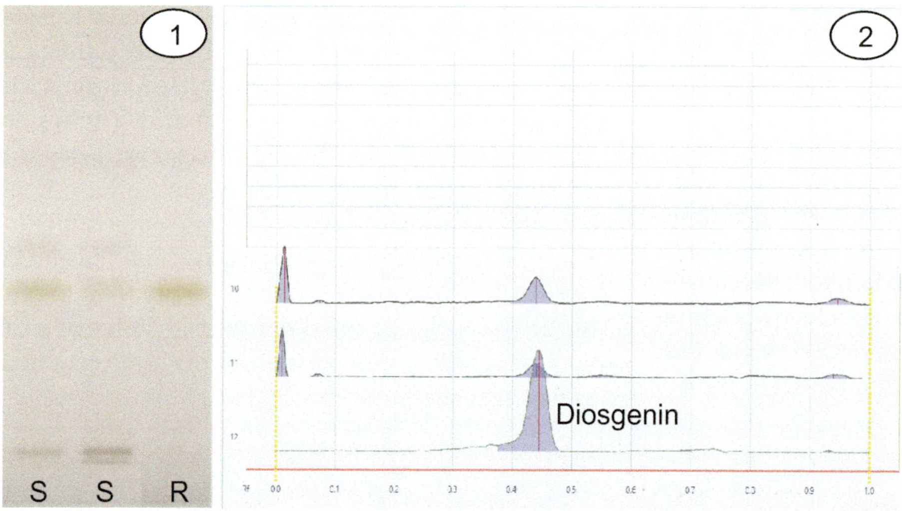

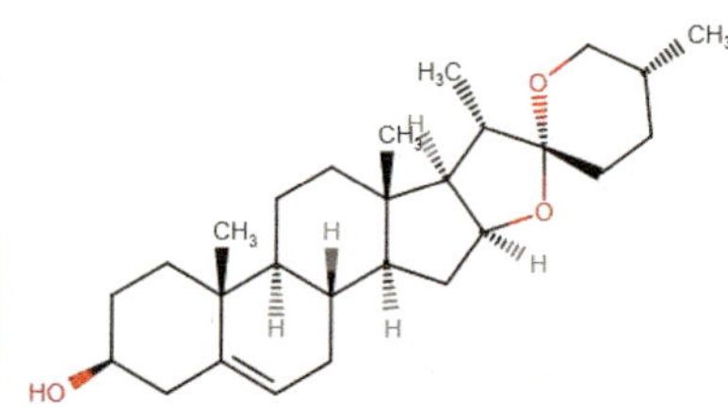

Diosgenin

Plate 1: HPTLC Chromatogram of *Diascorea deltoidea* Sample (S) and Standard Diosgenin (R).
Plate 2: Overlay Spectra of Sample and Standard.

1. Rokaya MB, Sharma LN (2016). *Dioscorea deltoidea* in Nepal: Cross Validating Uses and Ethnopharmacological Relevance. Asi J Ethnophar Med Food 2(2): 17-26.
2. Ali M (2012). Plant Growth - A Note on the Vegetative Growth in *Dioscorea deltoidea*. J Aroma Med plants 1(8): 2-6.
3. Gopichand, Singh RD, Meena RL, Kaul VK, Singh B (2013). Influence of manure and plant spacing on growth and yield of *Dioscorea deltoidea* Wall: an Endangered Species. J Med Plant Studies 1(3): 184-190.
4. Hamayun M (2007). Traditional use of some medicinal plants of Swat Valley, Pakistan. Ind J Trad Knowl 6(4): 636-641.
5. Kumari P, Joshi GC, Tiwari LM (2012). Indigenous uses of Threatened ethno medicinal plants used to cure different diseases by ethnic peoples of Almora district of Western Himalaya. Int J Aayur Her Med 2(4): 661-678.
6. Gao H, Hou B, Kuroyanagi M, Wu L (2007). Constituents from anti-tumorpromoting active part of *Dioscorea bulbifera* L. in JB6 mouse epidermal cells. Asian J Tradit Med 2: 104-109.
7. Selim S, Jaouni S Al (2015). Anticancer and apoptotic effects on cell proliferation of diosgenin isolated from *Costus speciosus* (Koen.) Sm. BMC Comple Alter Med 15(1): 301.
8. Yang CR, Zhang Y, Jacob MR, Khan SI, Zhang YJ, Li XC (2006). Antifungal activity of C-27 steroidal saponins. Antimicrob Agents Chemother 50(5): 1710–1714.
9. Itharat A, Houghton PJ, Eno-Amooquaye E, Burke PJ, Sampson JH, Raman A (2004). In vitro cytotoxic activity of Thai medicinal plants used traditionally to treat cancer. J Ethnopharmacol 90(1): 33–38.
10. Song FL, Gan RY, Zhang Y, Xiao Q, Kuang L (2010). Total phenolic contents and antioxidant capacities of selected chinese medicinal plants. Int J Mol Sci 11: 2362-2372.

28

Elephantopus scaber Linn.

Family

Asteraceae

Vernacular Name

Sanskrit: Prastarini, Kharaparnini; Hindi: Gobhi, Bantambakhu; Bengali: Hasti Pad; Marathi: Hastipata; English: Prickly leaved elephant's foot; Urdu: Gobhi; Telugu: Eddupattu; Malayalam: Koonjirikka.

Distribution [1]

Only this species of the genus has known to grown in India; it is distributed from Western Ghats, Chhattisgarh and Uttar Pradesh.

Traditional Uses [2–4]

Whole plant is used for the treatment of toothache. It is also used for the treatment of anemia, leucorrhea, asthma, diarrhea, cough and removing bladder stone.

Major Chemical Constituents [5–7]

Lupeol, deoxyelephantopin, isodeoxyelephantopin and ursolic acid.

Pharmacological Activity [8–11]

Antibacterial, antidiarrheal, Hepatoprotective and antioxidant activity.

Plant in Natural Habitat

Dried Root

Plant Description

It is a common weed growing up to 30 cm high, rigid, erect and hairy herb, mostly found in shady places. Stems are somehow woody and hairy; leaves are long, mostly basal rosette, oblong-ovate, pointed at the margin. Flowers are purple in colour and each flower consists about 4 flower, born in cluster at branch terminal, enclosed by three leaf like bract which also ovate to oblong-ovate and at the base. Roots are woody, cylindrical, adventitious, brownish white colored when dried.

Anatomy

Transverse section of mature root is almost circular in outline. The outer layer is made up of rectangular cork cells, followed by cork cambium. The cortical region is made up of round or irregular parenchymatous cells occupying major area of the section; larger towards central region. Single layer endodermis is clearly seen made of rectangular cells. The xylem is almost star shaped and present at the center, consists of vessels, fibers, tracheids and xylem parenchyma. Vessels are mostly in groups, and surrounded by parenchymatous phloem.

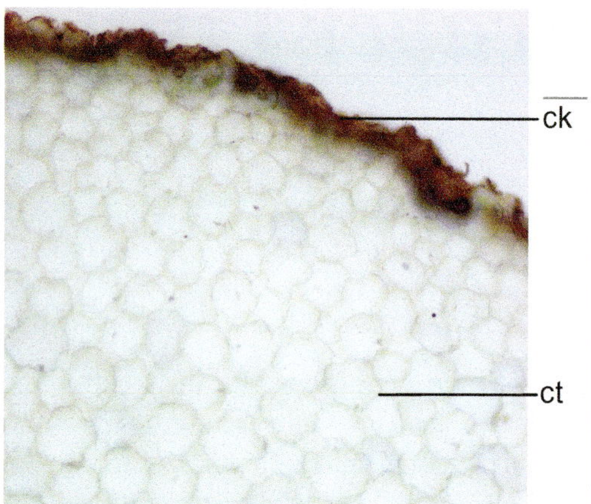

T.S. (20 x)

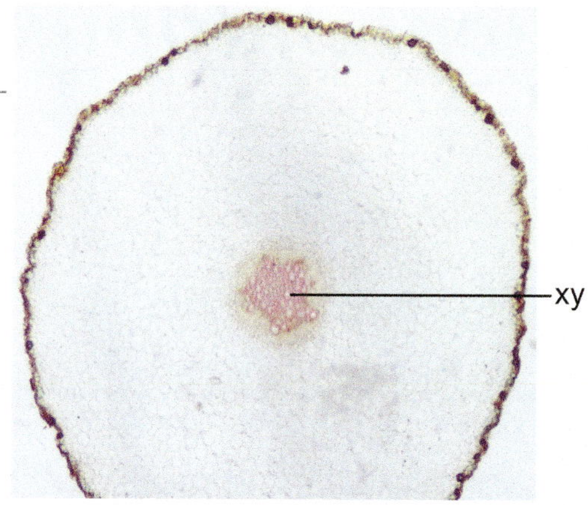

T.S. (4 x)

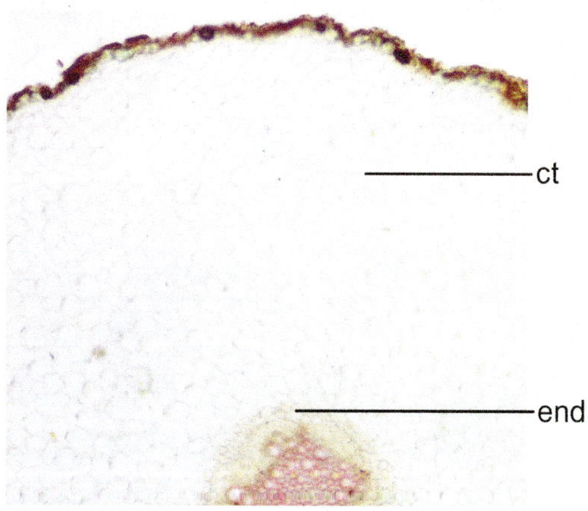

T.S. (10 x)

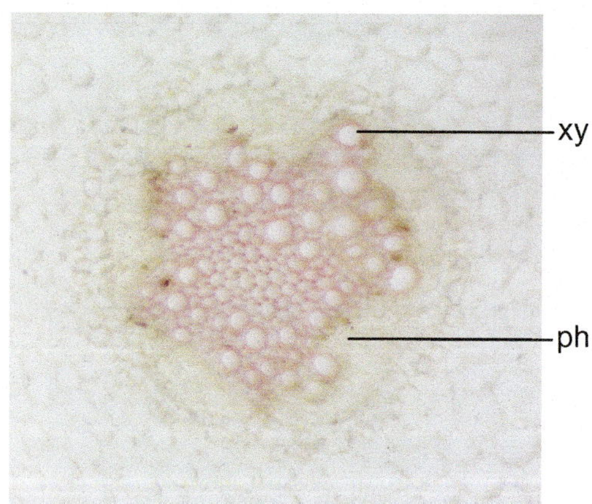

T.S. (20 x)

Transverse Section (T.S.) of *Elephantopus scaber* Root.

Abbreviation: ck- cork; ct- cortex; end- endodermis; ph- phloem; xy- xylem.

HPTLC Studies

Sample Solution

Accurately weighed (5g) Crude powdered sample was subjected to cold extraction with methanol (50 ml), continuously shaken for 6h and then allowed to stand at room temperature (25 ± 2^{0}C) for 18h. Extraction was repeated thrice, and the pooled filtrate was concentrate in rotatory evaporator (Buchi, Switzerland). Working solution of 10 mg/ml was prepared for HPTLC study.

Standard Solution

Lupeol (>98 per cent) was freshly prepared of strength 1 mg/ml.

Sample Application

2 µl of standard solution and 10 µl of sample was applied as 6 mm band with CAMAG HPTLC applicator on activated silica gel 60 GF 254 HPTLC plate. The chromatogram was developed on pre-saturated CAMAG

HPTLC twin trough chamber for 20 min and the solvent system was allowed to run for 8 cm.

Solvent System

Toluene- ethyl acetate- formic acid (7:3:0.5 v/v).

Visualization

The chromatogram was visualized after derivatization with anisaldehyde sulphuric acid reagent and densitometric scanning was done at 530 nm.

Identification of Lupeol

The standard lupeol was identified at R_f 0.82.

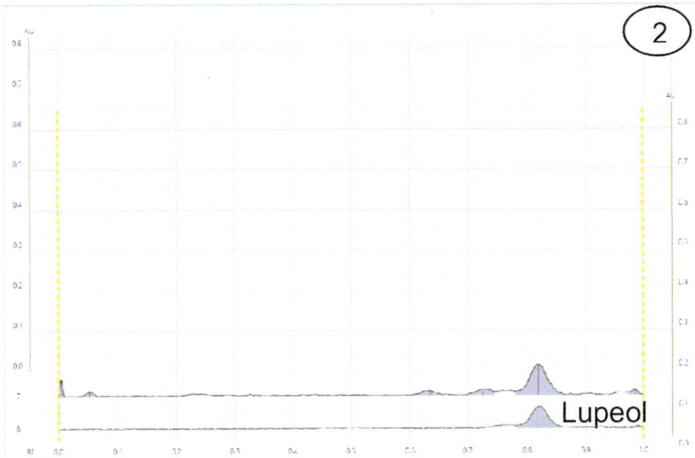

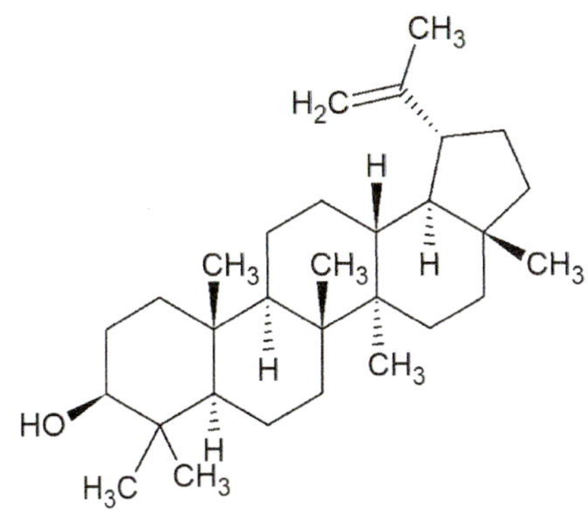

Lupeol

Plate 1: HPTLC Chromatogram of *Elephantopus scaber* (S) and Standard Lupeol (R), Visualized after Derivatization with Anisaldehyde Sulphuric Acid Reagent.

Plate 2: Overlay Spectra of Sample and Standard.

1. Geetha BS, Nair MS, Latha PG, Remani P (2012). Sesquiterpene lactones isolated from Elephantopus scaber l. inhibits human lymphocyte proliferation and the growth of tumour cell lines and induces apoptosis in vitro. J Biomed Biotechnol 1-8.
2. Achuta N, Shukla SS, Srivastava S, Rawat AKS (2010). An ethnobotanical study of medicinal plants of Rewa district Madhya Pradesh. Indian J traditional knowledge 9: 191-202.
3. Wijayakusuma H, Dalimartha S, Wirian AS, Yaputra T (1992). Medicinal Plants in Indonesia. Jakarta, Indonesia: Pustaka Kartini.
4. Poli A, Nicolau M, Simões CM, Nicolau RM, Zanin M (1992). Preliminary pharmacologic evaluation of crude whole plant extracts of *Elephantopus scaber*. Part I: in vivo studies. J ethnopharmacology 37(1):71-6.
5. Huang CC, Lo CP, Chiu CY, Shyur LF (2010). Deoxyelephantopin, a novel multifunctional agent, suppresses mammary tumor growth and lung metastasis and doubles survival time in mice. British journal of pharmacology 9(4): 856-71.
6. Liang QL, Gong ZN, Min ZD (2007). Studies on triterpenes from *E. scaber*. Chin Pharm J 42: 494-496.
7. Kiritikar KD, Basu BD (1991). Indian medicinal plants. 2nd ed. Dehradun: International Book Distributors 1328–1329.
8. Anitha VT, Antonisamy JM, Jeeva S (2012). Antibacterial studies on *Hemigraphis colorata* (Blume) H.G. Hallier and *Elephantopus scaber* L. Asian Pac J Trop Med 52-57.
9. Muthumani P, Chiristina AJM, Venkataraman S, Meera R, Devi P, Kameswari B, Eswarapriya B (2010). Anti-diarrhoeal and cardiotonic activity of extracts of *Elephantopus scaber* linn in experimental animals. Res J Pharmaceutical, Biological Chem Sci 1(3): 1.
10. Wan YH, Swee KY, Ho CL, Raha AR, Alitheen NB (2012). Hepatoprotective activity of *Elephantopus scaber* on alcohol-induced liver damage in mice. Evidence-Based Com Alt Med 1-20.
11. Sheeba KO, Wills PJ, Latha BK, Rajalekshmy R, Latha Ms (2012). Antioxidant and antihepatotoxic efficacy of methanolic extract of Elephantopus scaber Linn in Wistar rats. Asian Pacific J Trop Dis 2: 904-908.

29

Gloriosa superba Linn.

Family

Colchicaceae

Vernacular names [1-2]

Sanskrit: Kalikari, Agnisikha; Hindi: Kalihari, Languli; Bengali: Bishalanguli; English: Flame lily, Glory lily; Telugu: Kalappagadda, Ganjeri; Punjabi: Kariari; Tamil: Kizhangu; Malyalam: Mettoni.

Distribution [3-5]

Gloriosa superba is found throughout India up to an altitude of 6000 feet. It is native of tropical Africa and is now available in many parts of tropical Asia including India, Malaysia and Sri Lanka. The species is available in J and K, Uttar Pradesh, Madhya Pradesh, Chhattisgarh, Kerala, Haryana, Tamil Nadu, Karnataka, Sikkim and Orissa states of India.

Traditional Uses [6]

Gloriosa superba is an important species of Ayurvedic system of medicine. The tuber is traditionally used for the treatment of wounds, sprains, chronic ulcers, haemorrhoids and leprosy. The tubers are also used in the treatment of parasitic skin infections, leprosy and internal worms. Paste of tubers is applied to the lower part of the stomach to make the child's birth easier.

Major Chemical Constituents [7-8]

Colchicine and gloriosine

Major Pharmacological Activity [5, 9-11]

Anti-gout, anti-bacterial, anti-oxidant, anti-inflammatory and hepato-protective activity.

A Flowering Twig

Fresh Tubers

Plant Description

It is a perennial, tuberous, climbing herb with tendrils and easily propagated by tuber. Stem is soft; leaves are sessile, lanceolate, acuminate, glabrous, mainly alternate and sometime opposite. Flowers are axillary, solitary, actinomorphic, hermaphrodite; six tepals which are bright red to orange colour and yellowish base. Stamens 6, superior ovary, 3 celled with numerous ovules. Roots are brown, tuberous, smooth, cylindrical, fleshy, L or V shaped.

Anatomy

Transverse section of tuber is circular in outline, shows very narrow cork and wider cortex, occupying major area of section, well distinguished endodermal layer and wide stellar system. Detailed structure accounted that the rhizome have single layered epidermis, followed by multi-layered thin-walled cork cells; cortex and central ground tissues is parenchymatous, covers almost all the section and stealer vascular bundle is scattered throughout ground tissue. Steles are collateral, non-lignified fibro-vascular bundles; they are more numerous to peripheral region of endodermis. Stellar vascular bundles are more developed than cortical vascular bundle, composed of 4 to 5 vessels associated with narrow phloem. Simple oval to pear shaped starch grains are present throughout the ground tissue. In longitudinal section vessels are seen vertically running broad segmented canals, xylem fibres and phloem fibres are vertically arranged along with medullary rays.

HPTLC Studies

Sample Solution

Accurately weight (5 g) crude powdered sample was subjected to cold extraction with methanol (50 ml), continuously shaken for 6 h and then allowed to stand at room temperature (25 ^{0}C ± 2) for 18 h. The extraction was repeated thrice, filtered and the pooled filtrate was concentrate in rotatory evaporator (Buchi, Switzerland). Working solution of strength 10 mg/ml was freshly prepared on the day of quantative analysis.

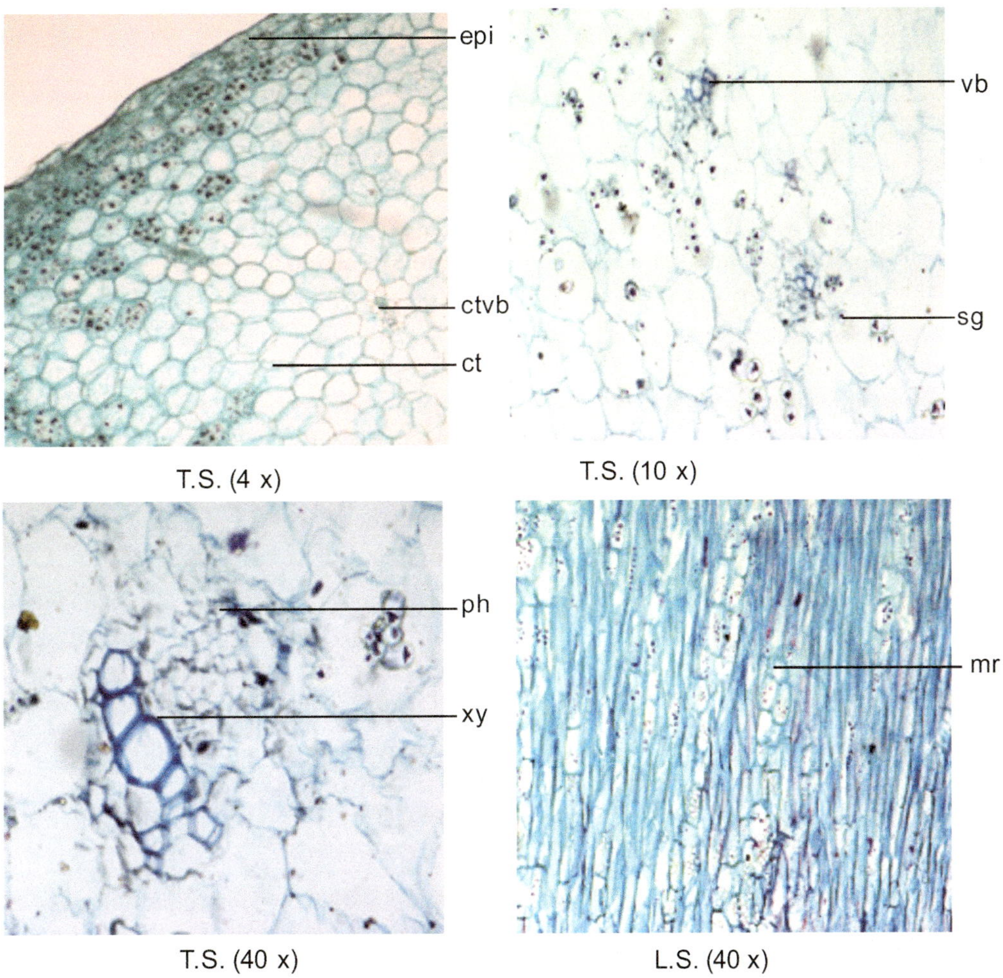

Transverse Section (T.S.) & Longitudinal Section (L.S) of *Gloriosa superba* Tuber.

Abbreviation: ck-cork; ct- cortex; ctvb- cortical vascular bundle; end- endodermis; mr- medullary rays; ph- phloem; sg- starch grains; vb- vascular bundle; ve- vessels; xy- xylem.

Standard Solution

Colchicine (≥ 98 per cent) was freshly prepared of strength 1 mg/ml.

Sample Application

2 µl of standard solution and 10 µl of sample was applied as 6 mm band with CAMAG HPTLC applicator on activated silica gel 60 GF_{254} HPTLC plate. The chromatogram was developed on pre-saturated CAMAG HPTLC twin trough chamber for 10 min and the solvent system was allowed to run for 8 cm.

Solvent System

Chloroform: acetone: diethyl amine (5:4:1 v/v)

Visualization

The chromatogram was visualized and densitometric scanning of colchicine was done at 350 nm.

Identification of Colchicine[12-18]

The standard colchicine was identified at R_f 0.52.

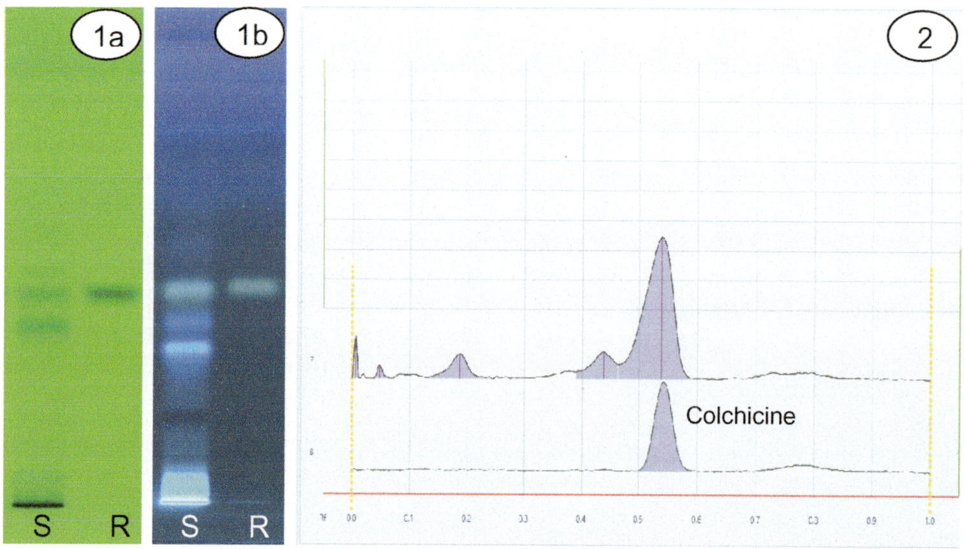

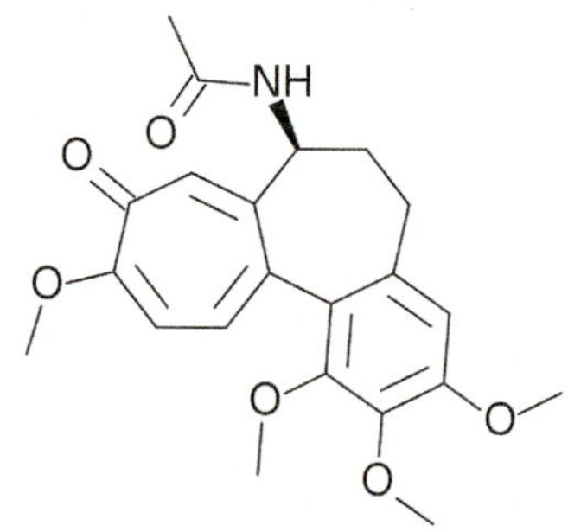

Colchicine

Plate 1: HPTLC Chromatogram of *Gloriosa superba* (S) and Reference Colchicine (R), 1a: Plate Visualized under UV_{254}; 1b: Plate Visualized under UV_{366}.
Plate 2: Overlay Spectra of Sample and Standard.

1. Pulliah T (2002). Medicinal plants in India, vol. I. New Delhi: Regency publication. 269-70.
2. Warrier PK, Nambier VPK, Ramankutty C (1995). Indian Medicinal Plant vol. 3 (Editor Arya Vaidya Sala, Kottakkal) Orient Longman. 76-81.
3. Ghosh B, Mukharjee S, Jha T B, Jha S (2002). Enhanced colchicine production in root cultures of *G. Superba* by direct and indirect precursors of the biosynthetic pathway. Biotechnol Lett 24: 231-4.
4. Singh AK (2006). Flowering crops: cultivation and management. New Delhi: New India publishing agency. 167-76.
5. Anonymous (1956). The wealth of India: a dictionary of raw material and industrial products, 4. New Delhi: CSIR. 139-40.
6. Jain SK (1991). Dictionary of Indian folk medicine and ethnobotany (a reference manual of man-plant relationship, ethnic groups and ethnobotany in India). New Delhi: Deep and Deep Publication.
7. Gooneratne BW (1966). Massive generalized alopecia after poisoning by *G. superba* Linn. Br Med J 231(5494): 1023-4.
8. Sarin YK, Jamwal PS, Gupta BK, Atal CK (1974). Colchicine from the seeds of *Gloriosa superba*. Curr sci 43: 87-90.
9. Hemaiswarya S, Raja R, Anbazhagan C, Thiagarajan V (2009). Antimicrobial and mutagenic properties of the root tubers of *Gloriosa superba* Linn. (Kalihari). Pak J Bot 41(1): 293-299.
10. Saradha M, Annapoorani S (2013). Antioxidative activity of methanolic extract of *Gloriosa superba* seed, tuber and leaves. Int J Pharma Res Develop 5(6): 102-108.
11. Gupta BK (1999). Production of colchicines from *G. superba* tubers in cultivation and utilization of medicinal plants. CSIR J: 270-8.
12. Ankita Misra, Priyanka Mishra, Bhanu Kumar, Pushpendra Kumar Shukla, Manish Kumar,

Satyendra Pratap Singh, Velusamay Sundaresan, Debyendhu Adhikari, Pawan Kumar Agrawal, Saroj Kanta Barik and Sharad Srivastava (2021). Intra-specific molecular and chemical variabilities in natural populations of *Gloriosa superba* (L.) for identifying elite chemotypes for commercial utilization, Fitoterapia. https://doi.org/10.1016/j.fitote.2021.104831

13. Ankita Misra, Mridul Kant Chaudhary, Pushpendra Shukla and Sharad Srivastava (2021). Simultaneous quantification of pharmacologically active alkaloid metabolites colchicine and gloriosine in *Gloriosa superba* collected from Western Ghats (India) for the identification of elite chemotype(s). J AOAC Intl. DOI: 10.1093/jaoacint/qsab007

14. Ankita Misra, Sharad Srivastava, Sanjeev Kumar, Pushpendra Kumar Shukla, Manish Kumar, Pawan Kumar Agrawal and Saroj Kanta Barik (2020): Chemotaxonomic studies on natural population of *Gloriosa superba* (L.) collected from Gangetic plain (India) and their *in vitro* antigout activity for the identification of elite germplasm(s). J Ethnopharmacol, https://doi.org/10.1016/j.jep.2019.112387

15. Ankita Misra, Bhanu Kumar, Pushpendra Shukla and Sharad Srivastava (2020). Simultaneous HPTLC-UV quantification of colchicine and gloriosine alkaloids in the natural population of *Gloriosa superba* L., collected from Eastern Ghats of India for the identification of elite chemotypes. J Liquid Chrom Rel Tech. 43(9–10): 351–360. DOI: 10.1080/10826076.2020.1725547

16. Ankita Misra, Sharad Srivastava, Pushpendra Kumar Shukla, Manish Kumar, Md. Khalid, Poonam Kushwaha and Pawan Kumar Agrawal (2019). Variability in alkaloid and phenolic content vis-a-vis antigout potential among the natural population of *Gloriosa superba* (L.) collected from Central India. Nat Prod Res DOI: 10.1080/14786419.2019.1675069

17. Ankita Misra, Akanksha Srivastava, Mohammed Khalid, Poonam Kushwaha, Sharad Srivastava (2017). Evaluation of anti arthritic potential of *Gloriosa superba* (L.) elite germplasm collected from Eastern Himalayas, India. Pharmacog J 9(6): 87-92.

18. Ankita Misra, Sharad Srivastava, Pushpendra Shukla, Jaichand and Ajay Kumar Singh Rawat (2017): High-performance Thin-layer Chromatographic-densitometric Quantification and Recovery of Bioactive Compounds for Identification of Elite Chemotypes of *Gloriosa superba* L. Collected from Sikkim Himalayas (India). Pharmacog Mag 13(51) S700-705.

30

Hemidesmus indicus (L.) R.Br.

Family

Apocynaceae

Vernacular name [1]

Sanskrit: Anantamula, Sariva, Nagajinva, Gopakanya; Hindi: Anantamul, Upalasari, Magrabu; Bengali: Anantmool; Marathi: Dudhasali; English: Indian Sarsaparilla; Gujarati: Sariva; Punjabi. :Anantmool; Telengu: Karibandha, Sogade.

Distribution [2]

Hemidesmus indicus is found throughout India, growing under semi dry conditions in the plains and up to an altitude of 600 m. It is also found in Sri Lanka, Pakistan, Iran and Bangladesh.

Traditional Uses [3–4]

In folk medicine, species is used to cure skin disease, inflammation, respiratory disorder, snake bite, fever, syphilis, asthma, kidney and in urinary disorder. It is also a good blood purifier.

Major Chemical Constituents [5]

Vanillin, Lupeol and β-sitosterol.

A Vegetative Plant

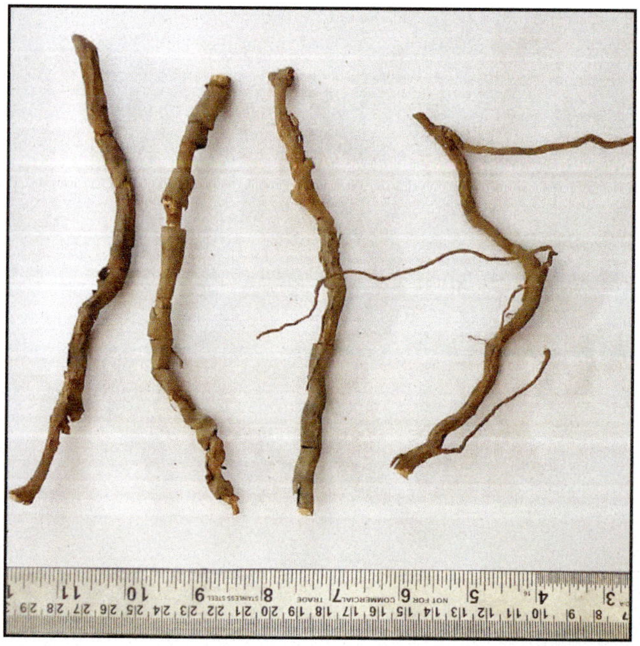

Dried Pieces of Root

Pharmacological Activity [6-10]

Anti-cancer, anti-ulcer, hepatoprotective, anti-diarrhoeal, anti-oxidant and anti-inflammatory activity.

Morphological Description

Hemidesmus indicus (L.) R.Br. is a perennial, lactiferous, twining, and prostrate and/or semi erect shrub. Roots are cylindrical dark brown woody and aromatic, the root length is more than 5 meters deep down to soil for this it called anantamul. Stem are numerous, brownish, cylindrical and thickening are located on nodes and the distance between two node is 5-8 cm. The leaves are simple, opposite, short petiole, dark green on ventral side and pubescent on dorsal. Flowers are greenish yellow to greenish purple, calyx five, corolla gamopetalous, five stamens, Pistil bicarpellary, and ovaries free, many ovules with distinct styles.

Anatomy

Transverse section of root is circular in outline, show very narrow range of cork and central wood, encircled by narrow phloem, wider cortex. Detailed structure accounted that the root shows narrow cork made of 2 to 3 row of dark brown cells, followed by wider cortex, cortical cells are parenchymatous ovoid and sometime circular and contains ovoidal starch grains. Pericycle is single layered. Vascular bundle is open type *i.e.* cambium presents between xylem and phloem; xylem are two types proto and metaxylem. Xylem vessels are thickened walled due to the deposition of lignin. Lignified medullary rays in continuation with phloem and radially arranged with xylem vessels, thin walled fibers and parenchyma. In longitudinal section vessels are seen vertically running broad segmented canals with bordered pitted and thickened wall, parenchyma are seen as narrow elongated tubes. The arrangements of medullary rays are horizontal, along with xylem vessels and fibers; tracheids are found along with the xylem vessels.

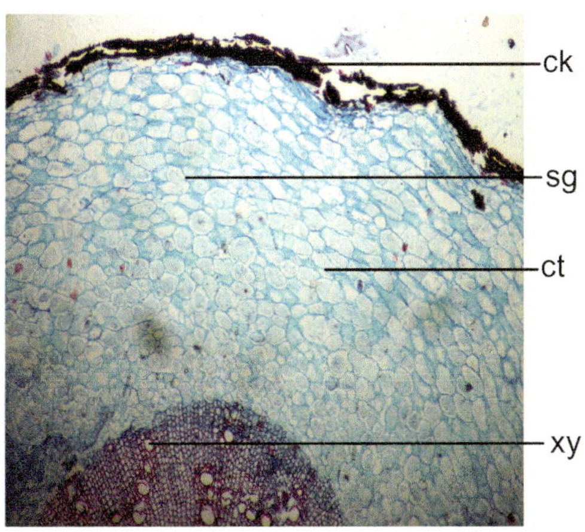

T.S. (4 x)

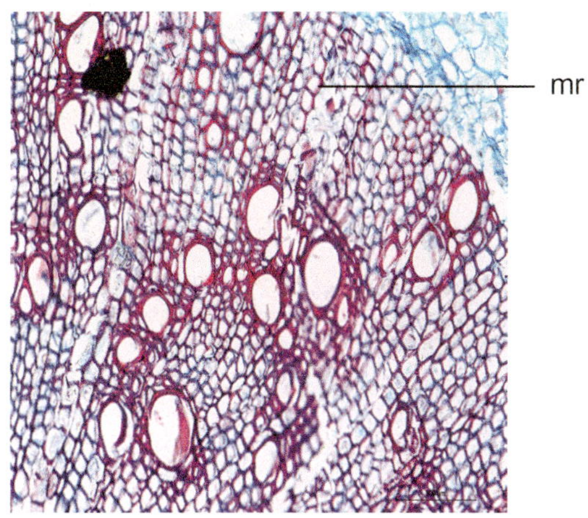

T.S. (20 x)

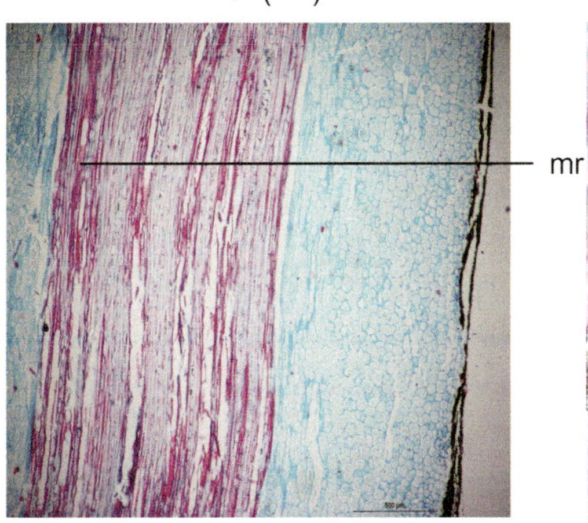

L.S. (4 x)

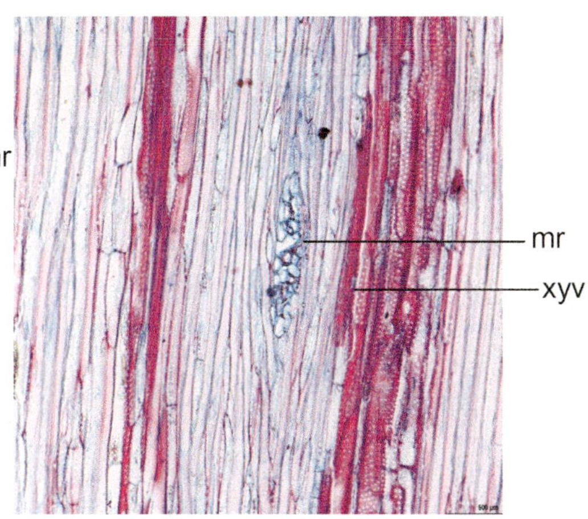

L.S (40 x)

Transverse Section (T.S) & Longitudinal Section (L.S.) of *Hemidesmus indicus* Root.

Abbreviation: ck- cork; ct- cortex; mr- medullary rays; sg- starch grains; xy- xylem; xyf-xylem fibre; xyv- xylem vessels.

HPTLC Studies

Sample Solution

Accurately weight (5 g) crude powdered sample was extracted (cold) with methanol (50 ml) followed by continuously shaken for 6h and then allowed to stand for 18 h at room temperature (25 ^{0}C ± 2). Extraction was repeated thrice, filtered and the pooled filtrate was concentrate in rotatory evaporator (Buchi, Switzerland), dried residue was stored at 4 ^{0}C. Working solution of strength 10 mg/ml was freshly prepared on the day of quantative analysis.

Standard Solution

Vanillin (≥ 98 per cent) was freshly prepared of strength 1 mg/ml.

Sample Application

2 µl of standard solution and 10 µl of sample was applied as 6 mm band with CAMAG HPTLC applicator on activated silica gel 60 GF_{254} HPTLC plate. The chromatogram was developed on pre-saturated Camag HPTLC twin trough chamber for 10 min and the solvent system was allowed to run for 8 cm.

Solvent System

n-hexane: Ethyl acetate (8:2 v/v)

Visualization

The chromatogram was visualized at UV254 and UV366 and densitometric scanning of vanillin was done at 320 nm.

Identification of Vanillin

The standard vanillin was identified at R_f 0.2

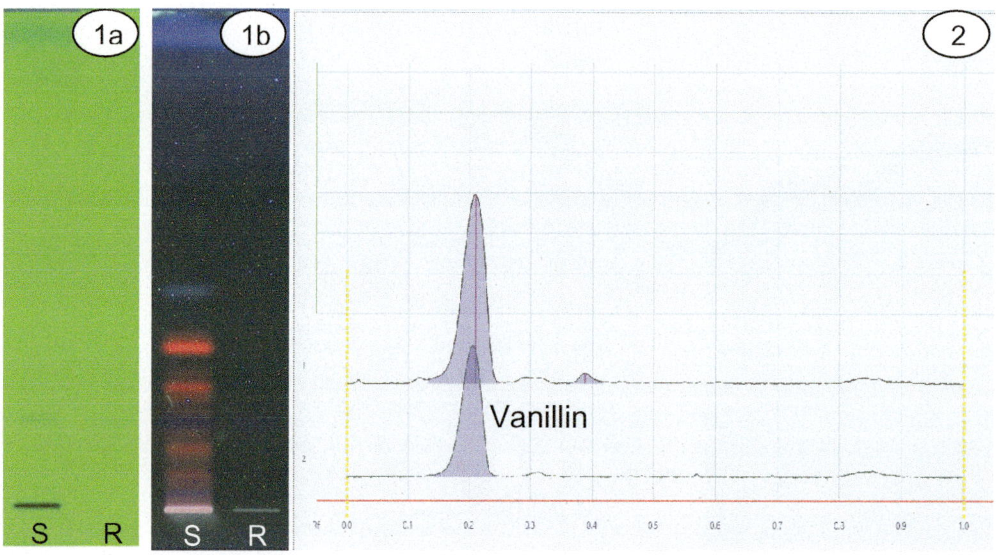

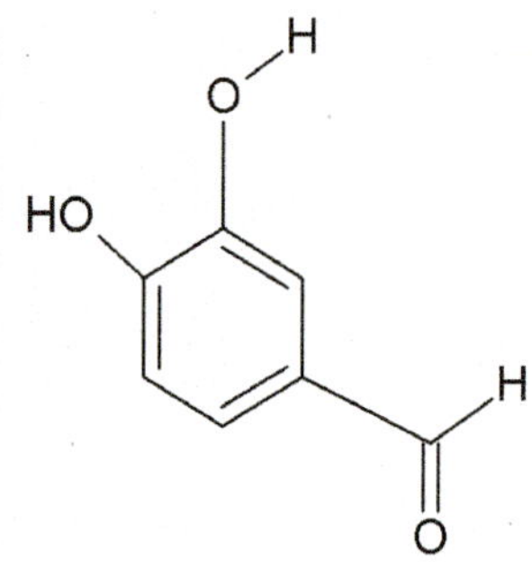

Vanillin

Plate 1: HPTLC Chromatogram of *Hemidesmus indicus* (S) and Reference Vanillin (R)
1a: Plate Visualized under UV254; 1: Plate Visualized under UV366.
Plate 2: Overlay Spectra of Sample and Standard.

1. Anonymous (2005). Quality Standards of Indian Medicinal Plants. Vol. 2. Indian Council of Medical Research, New Delhi 119-128.
2. Austin A (2008). A Review on Indian Sarsaparilla *Hemidesmus indicus* (L.) R.Br. J Biol sci 8(1): 1-12.
3. Mors WB (1991). Plants active against snake bite. Eco medici plant res 5: 353-73.
4. Nadkarni AN (1989). Indian material medica. Edn1, Vol 1, Popular Book Depot, Bombay, India 619.
5. Chandra R, Deepak D, Khare A (1994). Pregnane glycosides from *Hemidesmus Indicus*. Phytochemistry 35: 1545-8
6. Das S, Prakash R, Devaraj SN (2003). Anti-diarrhoeal effects of Methanolic root extract of *Hemidesmus indicus* (Indian Sarsaparilla) - an in vitro and in vivo study. Indian J Exp Biol 41: 363-6
7. Mary NK, Achuthan CR, Babu BH, Padikkala J (2003). In vitro Antioxidant and Antithrombotic Activity of *Hemidesmus indicus* (L.) R. Br., J Ethnopharmacol, August; 87(2-3): 187-91.

8. Nayar TS, Beegam AR, Mohanan N, Rajkumar G (2006). Flowering Plants of Kerala, A Handbook. Tropical Botanic Garden and Research Institute, Thiruvanathapuram, Kerala.

9. Prabakan M, Anandam R, Devaki T (2000). Protective effect of *Hemidesmus Indicus* against Rifampicin and Isoniazid-induced hepatotoxicity in rats. Fitoterapia 71: 55-9

10. Vijayalakshmi K, Shyamala R, Thirumurugan V, Sethuraman M, Rajan S, Badami S, Mukherjee PK (2010). Physico-phytochemical Investigation and Antiinflammatory Screening of *Capsicum annum* L. and *Hemidesmus indicus* (Linn.) R. Br. Anc Sci Life 29(4): 35-40.

31

Ichnocarpus frutescens Linn.

Family

Apocynaceae

Vernacular name [1]

Sanskrit: Syamalata, Sariva; Hindi: Kalidudhi, Siamalata; Bengali: Syamalota; Marathi: Krishnasarwa, Kantebhouri; English: Black creeper; Kannada: Gorwiballi; Oriya: Bhotinoi.

Distribution [2]

It is a climber found throughout the India as a weed up to an altitude of 1200 meter. It is also found in China, Australia and Srilanka.

Traditional Uses [3-5]

Root is used as alterative, tonic, diuretic and is used as a substitute of sarsaparilla. Traditionally the tribal communities of India use it for the treatment of diabetes, skin diseases, epilepsy and gall bladder stone. It is also used as blood purifier.

Major Chemical Constituents [6-7]

2-hydroxy-4-methoxybenzaldehyde, ferulic acid, ursolic acid, β-sitosterol

Pharmacological Activity [8-11]

Anticonvulsant, antidiabetic, hepatoprotective, anti-inflammatory and antioxidant activity

Plant in Natural Habitat

Dried Root

Morphological Description

Ichnocarpus frutescens Linn. is a perennial, lactiferous, twining, woody creeper, prostrate and/or semi erect with rusty red appearance shrub. The leaves are simple, opposite, alternate, short petiole, dark green on ventral side and pubescent on dorsal. Stem are numerous, woody cylindrical, climber, rough, exudes white latex when injured. Flowers are small, complete, bisexual, white or greenish white, actinomorphic, pentamerous, calyx five, corolla five gamopetalous, five stamens, two carpels, superior ovary with parietal placentation. Roots are cylindrical, dark brown coloured, woody and aromatic, the root length is more than 5 meters deep down in the soil.

Anatomy

Transverse section of root is circular in outline, show very narrow range of cork and central wood, encircled by narrow phloem and wider cortex. Detailed structure accounted that the root shows narrow cork made of 2 to 3 row of dark brown cells, followed by single layer epidermis and wider cortex, cortical cells are parenchymatous ovoid and sometime circular and contains ovoidal or circular starch grains. Vascular bundle is well developed, xylem are two types proto and metaxylem. Xylem vessels are thickened walled due to the deposition of lignin. Medullary rays in continuation with phloem and radially arranged with xylem vessels, thin walled fibers and parenchymatous conjunctive tissues.

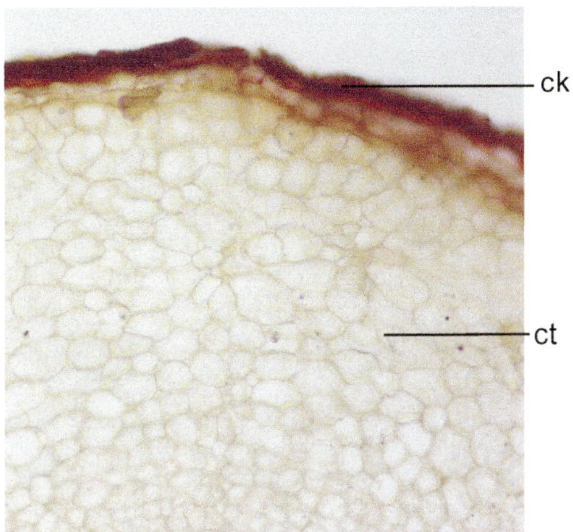

T.S. (10 x)

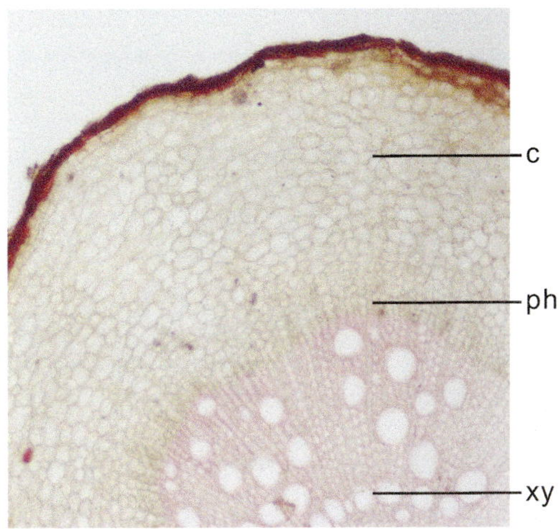

T.S. (4 x)

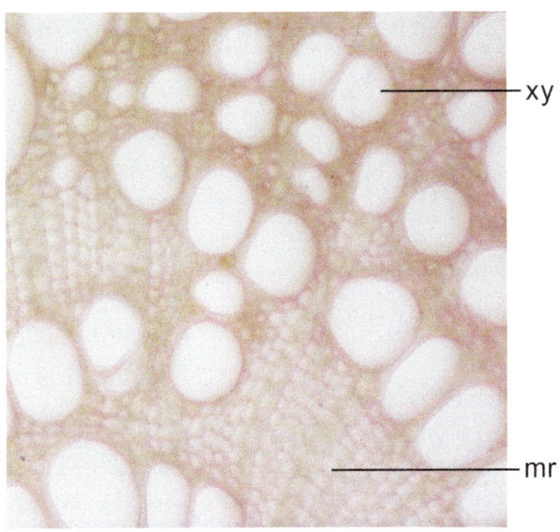

T.S. (20 x)

Transverse Section (T.S.) of *Ichnocarpus frutescence* Root.

Abbreviation: ck- cork; ct- cortex; mr- medullary rays; ph- phloem; xy- xylem.

HPTLC Studies

Sample Solution

Accurately weight (5 g) crude powdered sample was extracted (cold) with methanol (50 ml) followed by continuously shaken for 6h and then allowed to stand for 18 h at room temperature (25 ^{o}C ± 2). Extraction was repeated thrice, filtered and the pooled filtrate was concentrate in rotatory evaporator (Buchi, Switzerland), dried residue was stored at 4 ^{o}C. Working solution of strength 10 mg/ml was freshly prepared on the day of quantative analysis.

Standard Solution

2-hydroxy-4-methoxybenzaldehyde (≥98 per cent) was freshly prepared of strength 1 mg/ml.

Sample Application

2 µl of standard solution and 10 µl of sample was applied as 6 mm band with CAMAG HPTLC applicator on activated silica gel 60 GF_{254} HPTLC plate. The chromatogram was developed on pre-saturated Camag HPTLC twin trough chamber for 10 min and the solvent system was allowed to run for 8 cm.

Solvent System

n-hexane-ethyl acetate (8:2 v/v)

Visualization

The chromatogram was visualized in UV_{254} nm and densitometric scanning of 2-hydroxy-4-methoxybenzaldehyde was done at 320 nm.

Identification of 2-hydroxy-4-methoxybenzaldehyde

The standard vanillin was identified at Rf 0.2.

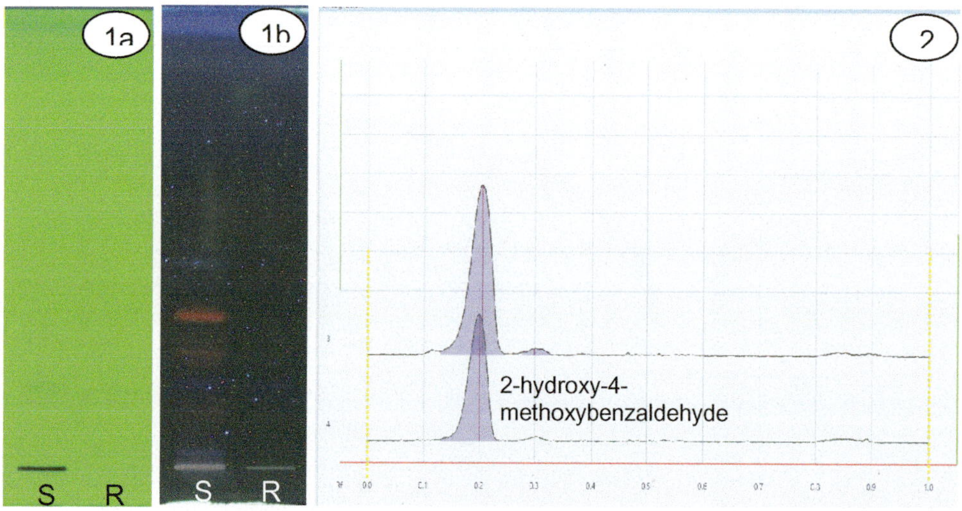

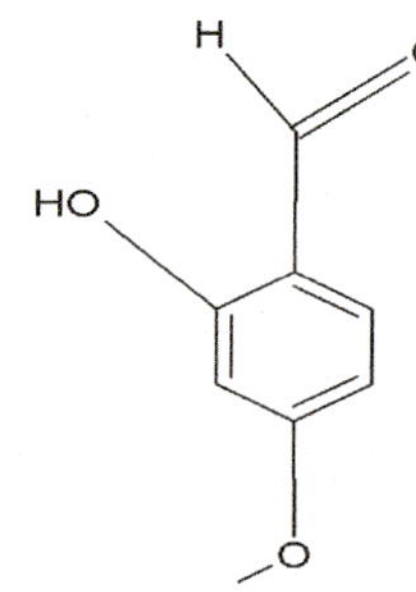

2-hydroxy-4-methoxybenzaldehyde

Plate 1: HPTLC Chromatogram of *Ichnocarpus frutescence* Sample (S) and Standard 2-hydroxy-4-methoxybenzaldehyde (R).

Plate 2: Overlay Spectra of Sample and Standard.

1. Ambasta S. P. (1999) 'Useful Plants of India', NISCOM, New Delhi, pp. 283
2. Medicinal Plants of India', Indian Council of Medical Research, New Delhi, Vol. 2 62-64, (1987).
3. Kirtikar KR, Basu BD (1999). Indian Medicinal Plants, Lalit Mohan Basu publication, Allahabad, India 1590-1592.
4. Rai MK (1988). Ehtnomedicinal survey of Patalkot and Tamiya (District Chhindwara): plants used against skin diseases and liver disorders. J Eco Taxo Bot 12: 337–339.
5. Goel AK, Mudgal V (1988). A survey of medicinal plants used by the tribals of Santal Pargana, Bihar. J Eco Taxo Bot 123: 29–335.
6. Pandurangan A, Khosa RL, Hemalatha S (2010). Chemical studies on the roots of *Ichnocarpus frutescens* [J]. Der Pharma Chemica 2(3): 222-224.
7. Chatterjee A. and Pakrashi S (2003). 'The Treatise of Indian Medicinal Plants', NISCAIR, New Delhi 4: 110-112.
8. Prathib B, Varghese NP (2016). Anticonvulsant Activity of Ethanol and Aqueous Extracts of *Ichnocarpus Frutescens* leaves in Experimental Mice. Int J Res Pharma Biosci 3(7): 1-6.
9. Pandurangan A, Khosa RL, Hemalatha S (2009). Evaluation of anti-inflammatory and antioxidant activity of *Ichnocarpus frutescens* root [J]. Daru-J Facul Pharm 17(1): 1-5.
10. Rakesh B, Sanjay J, Deep Q, Amit J, Girraj ST, Ravi G (2008). Anti-diabetic activity of aqueous root extract of *Ichnocarpus frutescens* in streptozotocin-nicotinamide induced typeII diabetes in rats. Indian J Pharmacol 40: 19-22.
11. Das DK, Yeligar VC, Nayak SS (2007). Evaluation of hepatoprotective and antioxidant activity of *Ichnocarpus frutescens* (Linn.) R. Br. on paracetamol-induced hepatotoxicity in rats [J]. Trop J Pharma Res 6(3): 755-765.

32

Mahonia borealis Takeda

Family

Berberidaceae

Vernacular name

Hindi: Gurang.

Distribution [1]

It is found in Uttarakhand, Himanchal Pradesh, Jammu and Kashmir (Western Himalaya) at an altitude of 1800 – 2700 m.

Traditional Uses [1-2]

Fruits are edible and root is used in treatment of diarrhea and dysentery. Root and stem are also used for yielding natural dye.

Major Chemical Constituents

Berberine

Pharmacological Activity [2-3]

Anti-proliferative, anti-migratory and anti-diabetic activity.

Plant Description

Mahonia borealis is an evergreen shrub, 1-2 m high and stem cylindrical, bark rough, withered and

A Flowering Twig

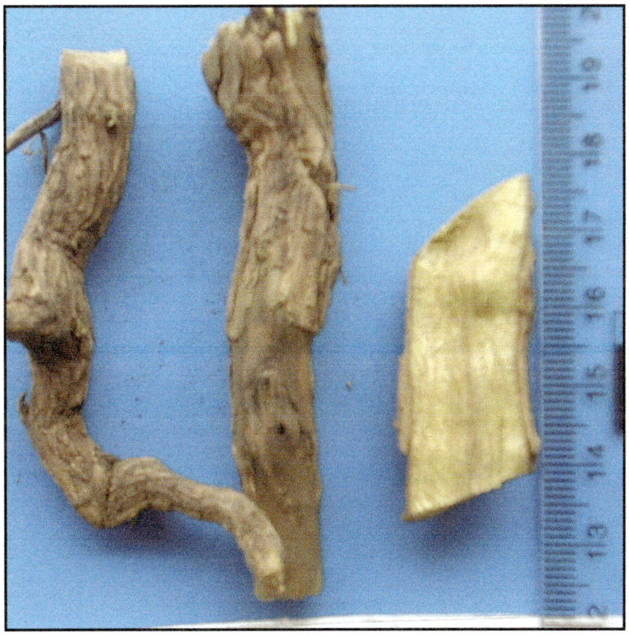

Dried Root

corky. Leaves are large, imparipinnate, opposite, 5 to 9 pairs leaflet with terminal one. Lower leaflet is small in comparison to upper, sessile or petiolulate, margin and apex are spiny dentate and prominent veins on both sides. Inflorescence is simple raceme, often with yellow flowers at the base. Fruit are berry, 7 - 8 mm long, 4 - 5 mm broad and globose or ovoid. Roots are hard, cylindrical in shape of 5 cm and the wood is pale yellow and distinctly radiated. The root bark is thick, externally corky and friable, internal smooth, closely striated and yellow in colour. Fracture is hard and bitter in taste.

Anatomy

The transverse section of root is circular in outline. The outermost zone is made of cork cell which are rectangular and 8 to 10 layered, followed by 2 to 3 layered cork cambium. Cortex is multilayered but narrowed due to secondary growth, filled with starch grains and sclereids. Sclereids are in groups of 4 to 10 or more and rarely solitary. The xylem is polyarch with 2 to 4 cells wide, heterogeneous, radiating medullary rays; most of the medullary rays are filled with starch gains. Xylem vessels are solitary or in groups of 1 or 2 and 1-2 layered vascular cambiums are also present. The central zone is narrow and mostly occupied by tracheids, xylem fibers and few vessels. In longitudinal section, medullary rays are heterogeneous and arrangements are horizontal, along with vessels and fibers; tracheids are present along with the xylem vessels.

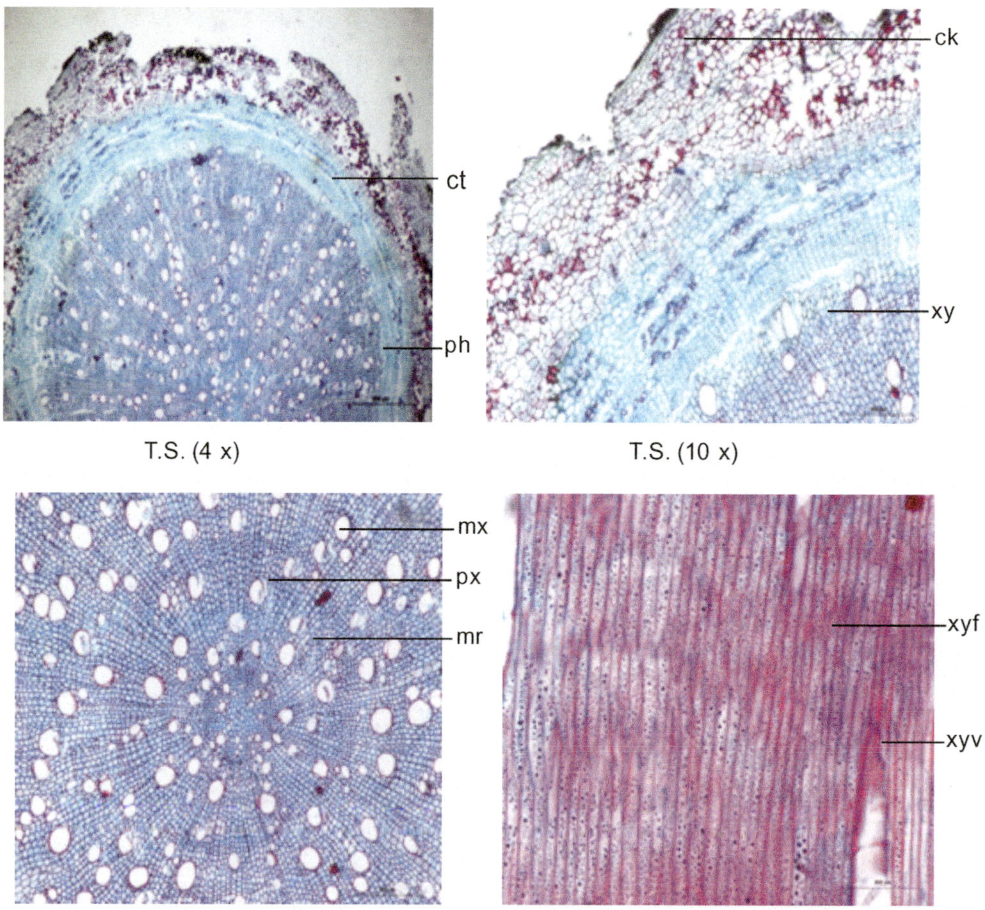

T.S. (4 x) T.S. (10 x) T.S. (20 x) L.S. (10 x)

Transverse Section (T.S.) & Longitudinal Section (L.S) of *Mahonia borealis* Root.

Abbreviation: ck- cork; ct- cortex; mr- medullary rays; mx- meta xylem; ph- phloem; px- proto xylem; xy- xylem; xyf- xylem fibre; xyv- xylem vessels.

HPTLC Studies

Sample Solution

Accurately weight (5 g) crude powdered sample was subjected to cold extraction with methanol (50 ml), continuously shaken for 6 h and then allowed to stand at room temperature (25 ^{0}C ± 2) for 18 h. The extraction was repeated thrice, filtered and the pooled filtrate was concentrate in rotatory evaporator (Buchi, Switzerland). Working solution of strength 10 mg/ml was freshly prepared on the day of quantative analysis.

Standard Solution

Berberine (≥ 98 per cent) was freshly prepared of strength 1 mg/ml.

Sample Application

2 µl of standard solution and 10 µl of sample was applied as 6 mm band with CAMAG HPTLC applicator on activated silica gel 60 GF_{254} HPTLC plate. The chromatogram was developed on pre-saturated CAMAG HPTLC twin trough chamber for 20 min and the solvent system was allowed to run for 8 cm.

Solvent System

n-propanol- water- formic acid (9:1:0.1 v/v)

Visualization

The chromatogram was visualized at UV_{366} and densitometric scanning of berberine was done at 350 nm.

Identification of Berberine

The standard berberine was identified at R_f 0.32

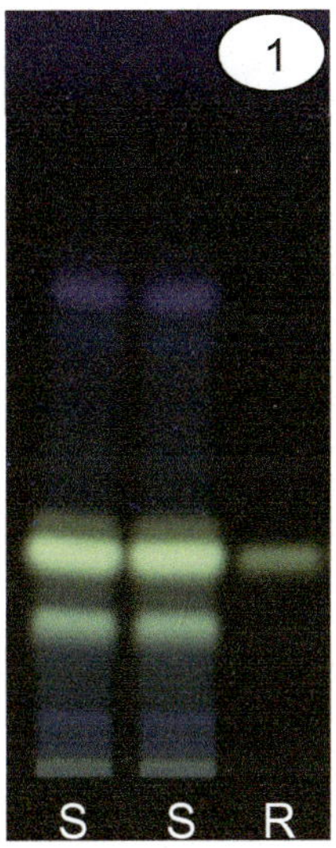

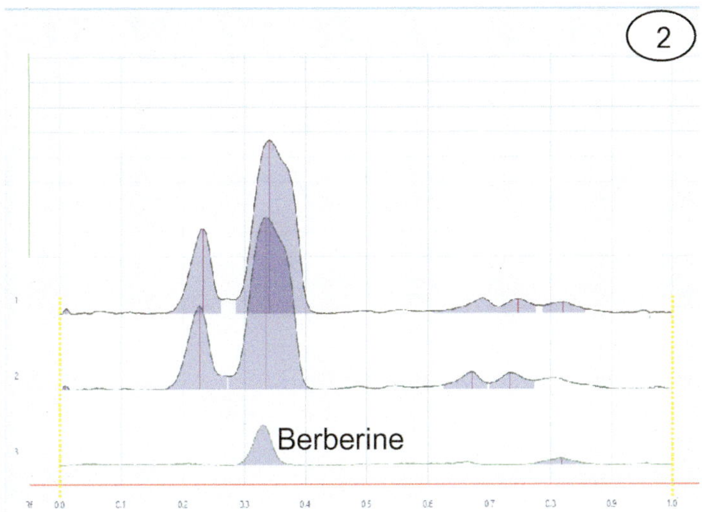

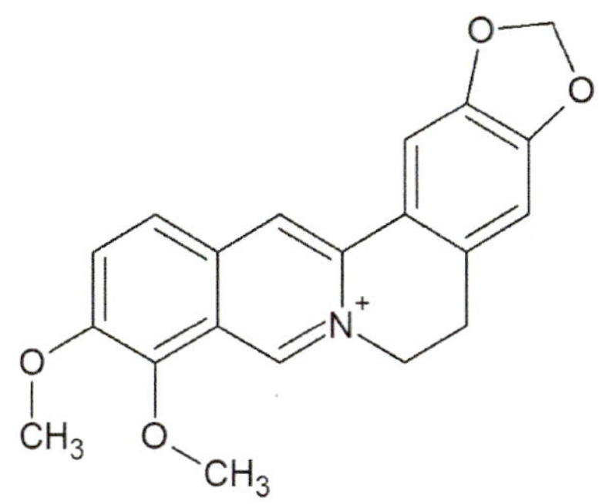

Berberine

Plate 1: HPTLC Chromatogram of *Mahonia borealis* Sample (S) and Standard Berberine (R).
Plate 2: Overlay Spectra of Sample and Standard.

1. Gaur RD (1999). Flora of district Garhwal, North Wes Himalaya, pp. 84. Trans Media, Srinagar, Garhwal, India.
2. Rohrer U, Kunz EM, Lenkeit K, Schaffner W, Meyer J (2007). Antimicrobial activity of *Mahonia aquifolium* and two of its alkaloids against oral bacteria. Schweiz Monatsschr Zahnmed; 117: 11261131.
3. Liang KW, Yin SC, Tingh CT, Lin SJ, Hsuesh CM, Chen C Y, Hsu SL (2008). Berberine inhibits platelet-derived growth factor-induced growth and migration partly through an AMPK-dependent pathway in vascular smooth muscle cells. Eur J Pharmacol 590:343-54.

33

Mahonia leschenaultii (Wallich ex Wight and Arn.) Takeda

Family

Berberidaceae

Vernacular name

English: Holy leaved Berry; Toda's term: Thovari.

Distribution [1]

It is found in Western Ghats, Nilgiri and Pulney hills at an altitude of 6000 feet.

Traditional Uses [2]

Paste of stem bark is used by the Todas in fever, cold, jaundice and postnatal treatment of women.

Major Chemical Constituents

Berberine

Pharmacological Activity [3–4]

Antimicrobial and anticancer activity

A Flowering Stalk

Dried Piece of Root

Plant Description

Mahonia leschenaultii is an evergreen shrub, attaining 1-2 m height. Stem is woody cylindrical, yellowish covered by rough bark, withered and corky. Twigs are glabrous, pale yellowish; leaves are oblong, elliptic or broadly obovate and dentate margin, dark green with prominent primary and secondary reticulate venation. Inflorescence is simple raceme, often with yellow flowers at the base. Fruit are 7-10 mm long, ovoid, blue-black with distinct style. Roots are hard, cylindrical in shape of 5 cm (approx.) or more in diameter. The wood is pale yellow and distinctly radiated, bark is thick, externally corky and friable. Internally it is smooth, closely striated and yellow in colour. Fracture is hard and bitter in taste.

Anatomy

The transverse section of root is circular in outline. The outermost zone is made of cork cell which are rectangular and 10 to 14 layered, followed by 1 to 2 layered cork cambium. Cortex, 15 to 20 layered, narrowed due to secondary growths; filled with starch grains and sclereids. Sclereids are in groups of 2 to 10 or more and rarely solitary. The xylem is polyarch, with 2 to 4 cells wide, vessels are solitary or in groups of 2 or 5. The central zone is narrow and mostly occupied by tracheids, xylem fibers and few vessels. Medullary rays are heterogeneous, multiseriate radiating, and contain starch grains. In longitudinal section, arrangements of medullary rays are horizontal, along with vessels and fibers; tracheids are found along with the xylem vessels.

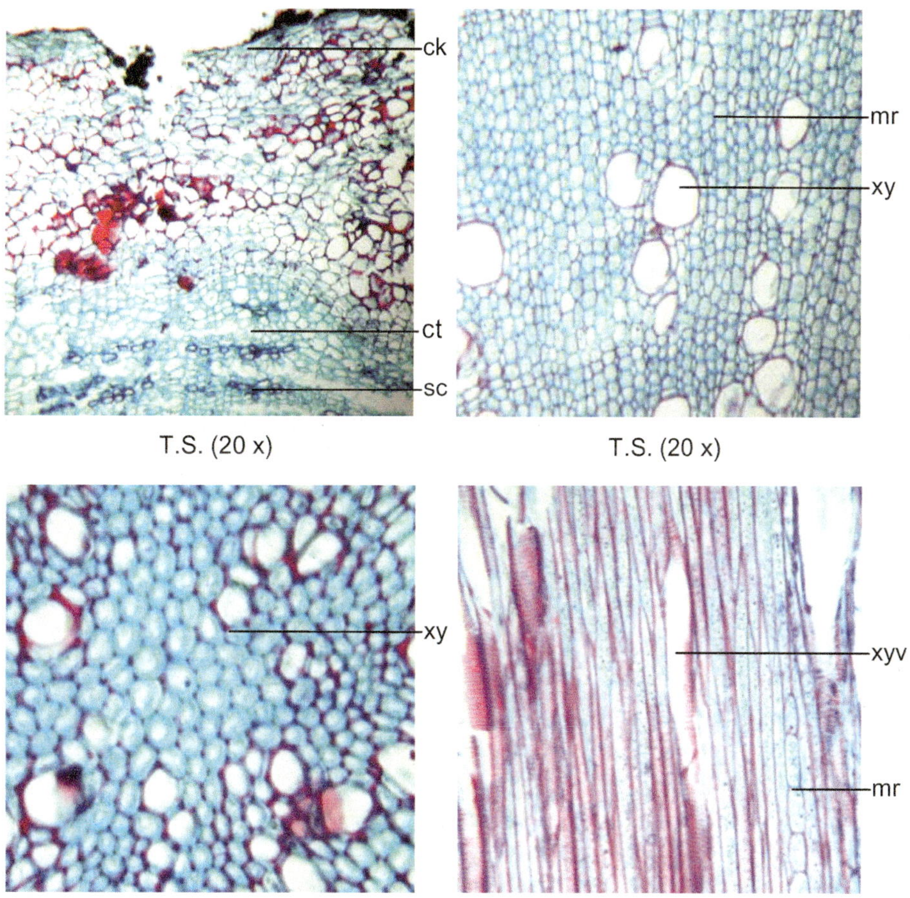

T.S. (20 x) T.S. (20 x) T.S. (20 x) L.S. (40 x)

Transverse Section (T.S.) & Longitudinal Section (L.S) of *Mahonia leschenaultii* Root.
Abbreviation: ck- cork; ct- cortex; mr- medullary rays; sc- sclereids; xy- xylem; xyv- xylem vessels.

HPTLC Studies

Working Solution

Accurately weight (5 g) crude powdered sample was subjected to cold extraction with methanol (50 ml), continuously shaken for 6 h and then allowed to stand at room temperature (25 ^{0}C ± 2) for 18 h. The extraction was repeated thrice, filtered and the pooled filtrate was concentrate in rotatory evaporator (Buchi, Switzerland). Working solution of strength 10 mg/ml was freshly prepared on the day of quantative analysis.

Standard Solution

Berberine (≥ 98 per cent) was freshly prepared of strength 1 mg/ml.

Sample Application

2 µl of standard solution and 10 µl of sample was applied as 6 mm band with CAMAG HPTLC applicator on activated silica gel 60 GF_{254} HPTLC plate. The chromatogram was developed on pre-saturated CAMAG HPTLC twin trough chamber for 30 min and the solvent system was allowed to run for 8 cm.

Solvent System

n-propanol: water: formic acid (9:1:0.1 v/v)

Visualization

The chromatogram was visualized at UV_{366} and densitometric scanning of berberine was done at 350 nm.

Identification of Berberine

The standard berberine was identified at R_f 0.32

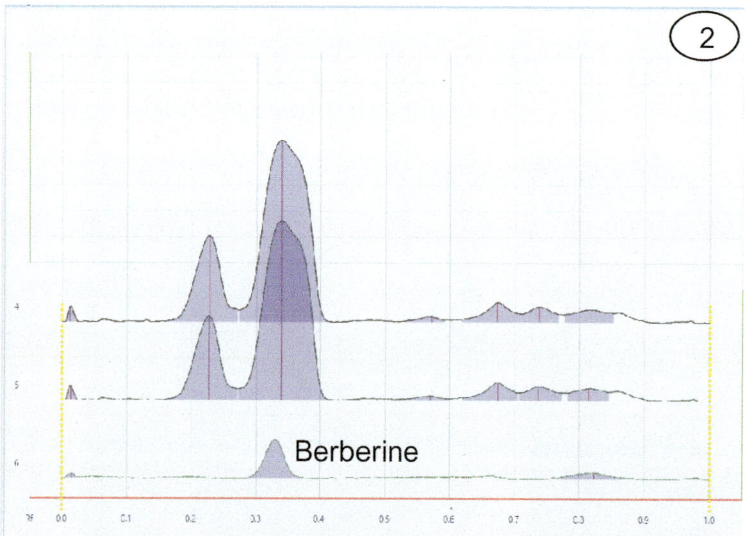

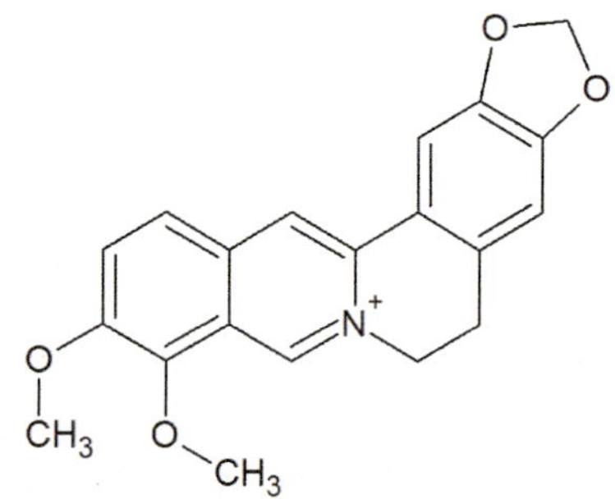

Berberine

Plate 1: HPTLC Chromatogram of *Mahonia leschenaultii* Sample (S) and Standard Berberine (R).
Plate 2: Overlay Spectra of Sample and Standard.

1. Gamble JS (1984). "Flora of the Presidency of Madras" 1st edition, Vol. I, Bishen Shing Mahendra Pal Singh, Dehradun: 32.
2. Nair NC, Henry AN (1983). Flora of Tamil Nadu, Series 1: Analysis Vol.1 Botanical Survey of India, Howrah.
3. Nitha B, Remashree AB, Balachandran I (2012). Antibacterial activity of some selected Indian medicinal plants. Int J Pharma Sci Res 3(7): 20-38.
4. Latha R, Rajanathan TC, Khusro A, Chidambaranathan N, Agastian P, Nagarajan S (2019). Anticancer activity of *Mahonia leschenaultii* methanolic root extract and berberine on Dalton's ascitic lymphoma in mice. Asi Paci J Tropi Med 12(6): 264.

34

Mahonia nepaulensis DC.

Family

Berberidaceae

Vernacular name

Nepali: Jamanemandro; Newari: Michiki swan

Distribution

The species is native of Nepal and is widely distributed in the mountains, at an altitude of about 1000-2000 m in India, Nepal, Bhutan, China and Vietnam

Traditional Uses

Traditionally it is used for treatment of skin problems such as psoriasis, dermatitis and eczema.

Major Chemical Constituents [1-2]

Berberine and jatrorrhizine

Pharmacological Activity [3-4]

Anti-microbial and anti-diabetic activity

Plant Description

Mahonia nepaulensis is large deciduous shrubs, twigs yellowish brown, hard, erect, cylindrical, smooth

A Leafy Twig

Dried Root

and bright yellow. Leaves are also large, dark green, imparipinnate, opposite, leaflets are elliptic to ovate, sessile or petiolulate, 4 to 11 pairs leaflet with terminal one. Margin and apex of leaflets are prominently spiny dentate, terminal leaflets are larger in comparison of lateral. Inflorescence is fascicled racemes approximately 20 cm long, with many tiny yellow flowers. Fruits are small berry blue or black colored. Roots are thick, woody, yellowish, cylindrical, covered with a thin friable bark. Bark is internally yellowish, rough, fracture hard and odorless.

Anatomy

Transverse section of mature root is circular in outline. The outer layer is made up of rectangular 8 to 11 layered cork cells, followed by 2 or 3 layered cork cambium. The cortical region is multilayered irregular parenchymatous cells, larger towards central region and contains sclereids. Sclereids are mostly solitary or in a group of 4 to 10 or more. Just below the cortex 4 to 10 layered thick phloem having sieve tubes, companion cells and phloem fibers are present. The xylem is 4-6 celled wide and consists of vessels, fibers, tracheids and xylem parenchyma. Vessels are mostly solitary or in groups of 2 to 5. The medullary rays are thick-walled, radiating, and parenchymatous. In longitudinal section, arrangements of medullary rays are horizontal, along with vessels and fibers; tracheids are found along with the xylem vessels.

HPTLC Studies

Sample Solution

Accurately weight (5 g) crude powdered sample was subjected to cold extraction with methanol (50 ml), continuously shaken for 6 h and then allowed to stand at room temperature (25 °C ± 2) for 18 h. The extraction was repeated thrice, filtered and the pooled filtrate was concentrate in rotatory evaporator (Buchi, Switzerland). Working solution of strength 10 mg/ml was freshly prepared on the day of quantative analysis.

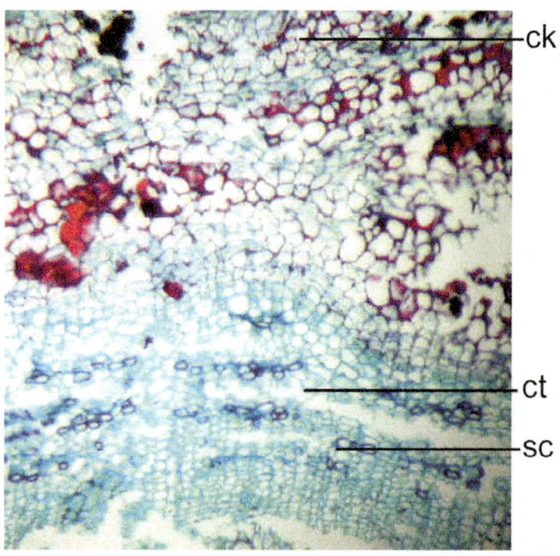

T.S. (40 x)

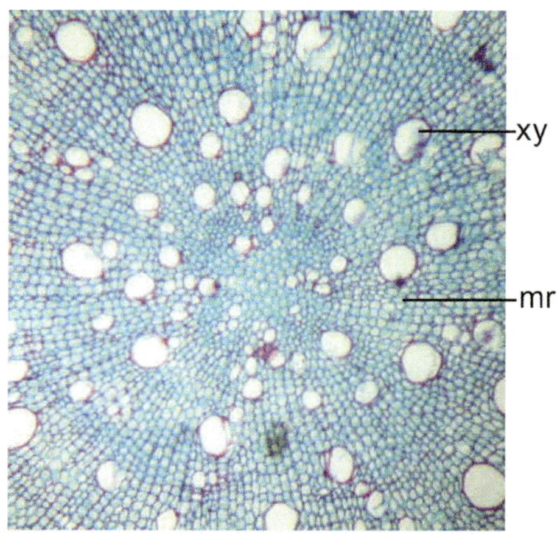

T.S. (20 x)

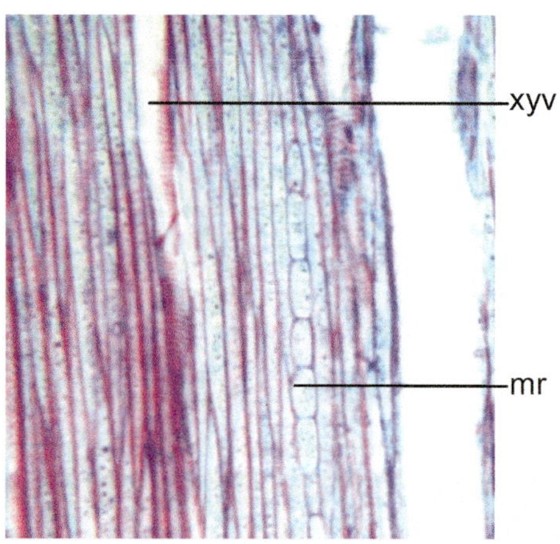

L.S. (40 x)

Transverse Section (T.S.) & Longitudinal Section (L.S) of *Mahonia nepaulensis* Root.
Abbreviation: ck- cork; ct- cortex; mr- medullary rays; sc- sclereids; xy- xylem; xyv- xylem vessels.

Standard Solution

Berberine (≥ 98 per cent) was freshly prepared of strength 1 mg/ml.

Sample Application

2 μl of standard solution and 10 μl of sample was applied as 6 mm band with CAMAG HPTLC applicator on activated silica gel 60 GF_{254} HPTLC plate. The chromatogram was developed on pre-saturated CAMAG HPTLC twin trough chamber for 30 min and the solvent system was allowed to run for 8 cm.

Solvent System

n-propanol: water: formic acid (9:1:0.1 v/v)

Visualization

The chromatogram was visualized at UV_{366} and densitometric scanning of berberine was done at 350 nm.

Identification of Berberine

The standard berberine was identified at R_f 0.32.

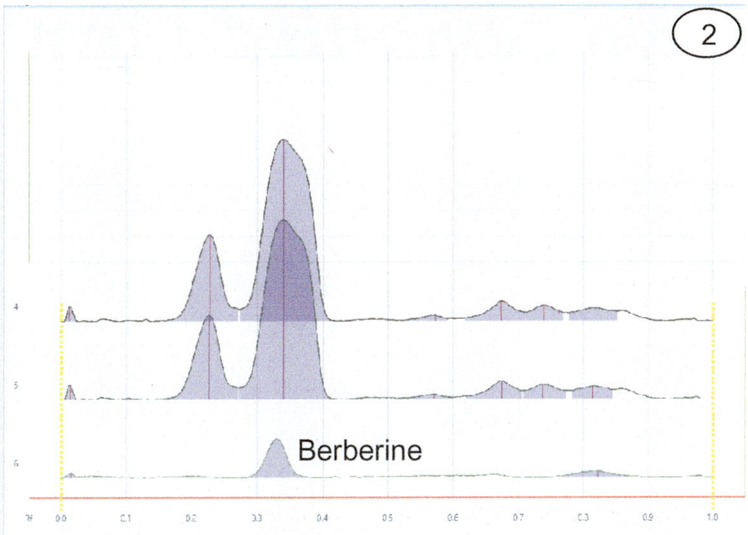

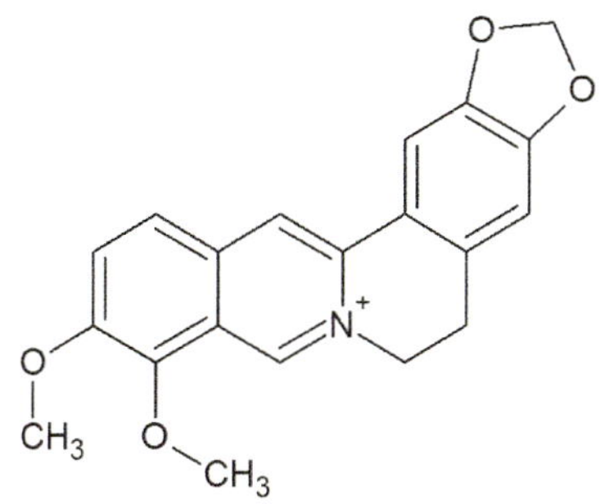

Berberine

Plate 1: HPTLC Chromatogram of *Mahonia nepaulensis* Sample (S) and Standard Berberine (R).
Plate 2: Overlay Spectra of Sample and Standard.

1. Govindachari TR, Pai BR, Rajadurai, S, Rao UR (1957). Alkaloids of Mahonia nepalensis D.C. J Ind Chem Soc: 41-48.
2. Nguyen TM, Tran AT, Hoang TH, Chau VM, Ninh KB, Phan VK (2007). Secobisbenzylisoquinoline Alkaloid from Mahonia nepalensis D.C. J Chem 46(5): 63-68.
3. Thusa R, Mulmi S (2017). Analysis of Phytoconstituents and Biological Activities of Different Parts of Mahonia nepalensis and Berberis aristata. Nep J Biotec 5(1): 5-13.
4. Das A, Chhetry TK (2016). Evaluation of Anti-Diabetic Activity of Mahonia nepalensis in STZ Induced Rat Model. Int J Pharmacog Phytochem Res 8(7): 1104-1110.

35

Plumbago zeylanica Linn.

Family

Plumbaginaceae

Vernacular Names [1]

Sanskrit: Chitra; Hindi: Chira, Chitrak; Bengali: chitra; English: Lead wort; Gujarati: Agni; Punjabi: Chitra; Telugu: chitramoolam; Malyalam: Veellakeduveli.

Distribution [2-3]

Plumbago zeylanica is found throughout India ranging from Central India to West Bengal, Maharashtra, Uttar Pradesh and in parts of South India. The species spreads across the World's tropical and sub-tropical countries, growing in deciduous woodland, savannahs and scrublands up to 2000 m altitude.

Traditional Uses [4-6]

In Ayurvedic and Unani system of medicines the entire plant is used in anemia, rheumatic pain, dysmenorrhoea, scabies, leprosy, inflammation and elimination of intestinal parasites. Roots are used as anti-atherogenic, cardiotonic, neuroprotective and

A Flowering Twig

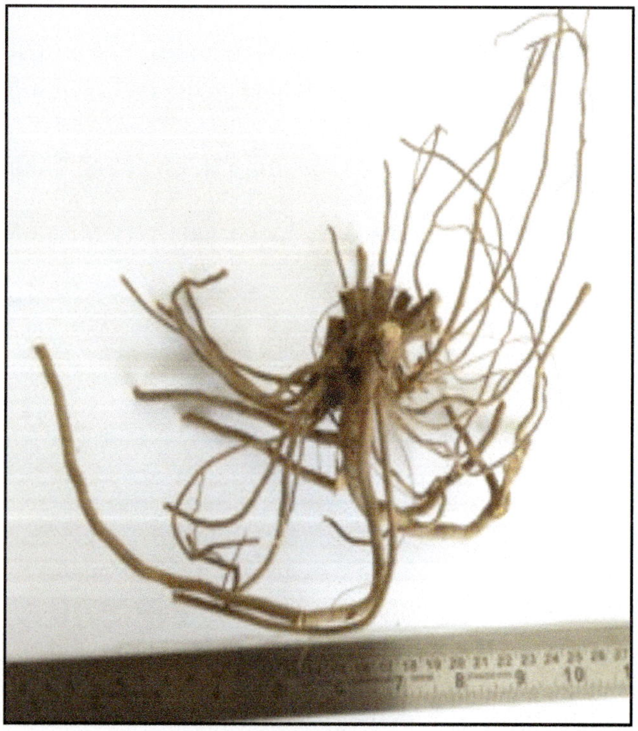

Dried Root

central nervous system stimulant. In Indian system of medicine the species is recommended for indigestion and appetizer, hemorrhoids, diarrhea and skin diseases.

Major Chemical Constituents [7-9]

Plumbagin, Chitranone, Seselin, 5-methoxyseselin, suberosin

Pharmacological Activity [10-14]

Anti-bacterial, anti-diabetic, cardiotonic, anti-convulasant, neuroprotective and anti-fertility.

Plant Description

Plumbago zeylanica L. is a herbaceous plant with glabrous stem that are prostrate or erect and sometime climber. It is perennial shrub attaining a height about 1-2 meters. Leaves are alternate, simple and ovate with hairy margin. Inflorescence is of terminal raceme type. Flowers are white in colour, pentamerous, small, axillary, bisexual, regular and in clustered form. Calyx are tubular with glandular trichome secretes sticky substance, corolla are five and white in colour. Style is filiform with five elongated stigma lobes and single celled superior ovary. Roots are seen yellowish when fresh and after dry it turns to reddish brown; it is cylindrical, smooth, straight long, sometime irregularly bent and with or without secondary root. Root has acrid odour and bitter taste.

Anatomy

Transverse section of root is circular in outline, shows narrow range of cork and central wood, encircled by narrow phloem and wider cortex. Detailed structure accounted that the root shows narrow cork made of 2 to 3 row of dark brown cells, followed by wide parenchymatous cortex, filled with reddish brown content and starch grains; pericycle shows discontinuous fibres; phloem narrow, parenchymatous, also filled with reddish brown content, cambium distinct. Lignified medullary rays in continuation with phloem filled with dark brown content and radially arranged with xylem vessels, thin walled fibres and parenchyma. In longitudinal section vessels are seen vertically running broad segmented canals with bordered pitted and thickened wall, parenchyma are seen as narrow elongated tubes with pitted walls. The arrangements of medullary rays are horizontal, along with xylem vessels and fibres.

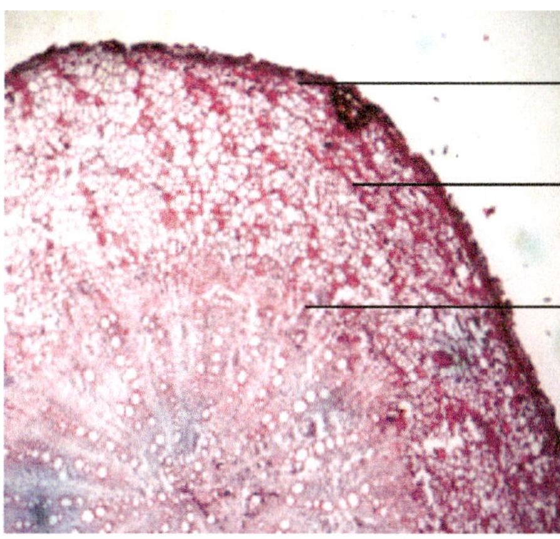

T.S. (4 x)

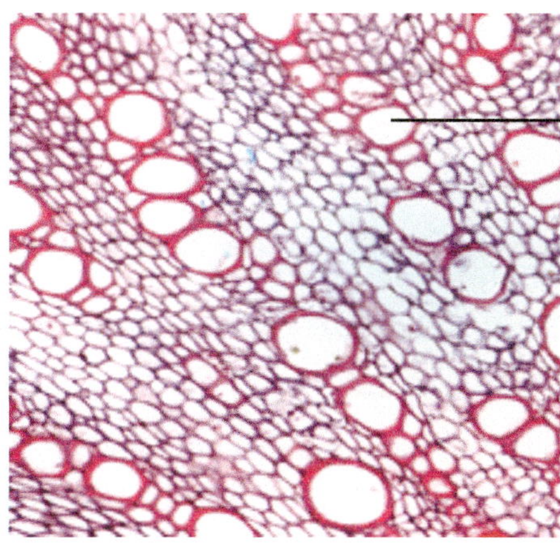

T.S. (20 x)

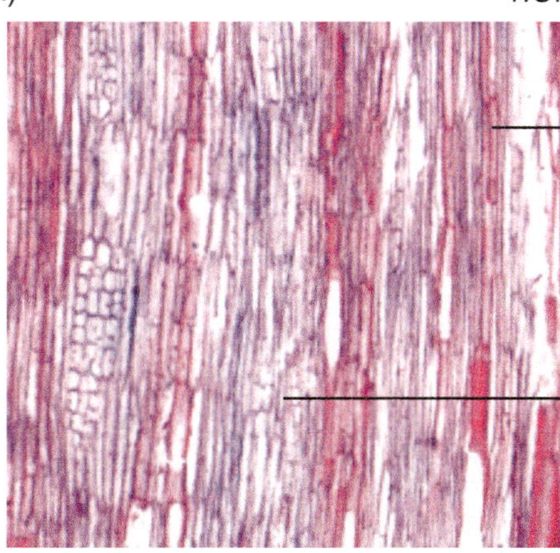

L.S. (20 x)

Transverse Section (T.S.) & Longitudinal Section (L.S) of *Plumbago zeylanica* Root.

Abbreviation: ck- cork; ct- cortex; mr- medullary rays; ph- phloem; ve- vessels; xy- xylem; xyv- xylem vessels.

HPTLC Studies

Sample Solution

Accurately weight (5 g) crude powdered root was extracted (cold) with methanol (50 ml) followed by continuously shaken for 6 h and then allowed to stand for 18 h at room temperature (25 ^{0}C ± 2). Extraction was repeated thrice, filtered and the pooled filtrate was concentrate in rotatory evaporator (Buchi, Switzerland), dried residue was stored at 4 ^{0}C. Working solution of strength 10 mg/ml was freshly prepared on the day of quantative analysis.

Standard Solution

Plumbagin (≥ 98 per cent) was freshly prepared of strength 1 mg/ml.

Sample Application

2 µl of standard solution and 10 µl of sample was applied as 6 mm band with CAMAG HPTLC applicator on activated silica gel 60 GF_{254} HPTLC plate. The chromatogram was developed on pre-saturated Camag HPTLC twin trough chamber for 10 min and the solvent system was allowed to run for 8 cm.

Solvent System

Toluene: Ethyl acetate (7:3 v/v)

Visualization

The chromatogram was visualized at UV_{254} and UV_{366} and densitometric scanning of Plumbagin was done at 270 nm.

Identification of Plumbagin

The standard Plumbagin was identified at R_f 0.61.

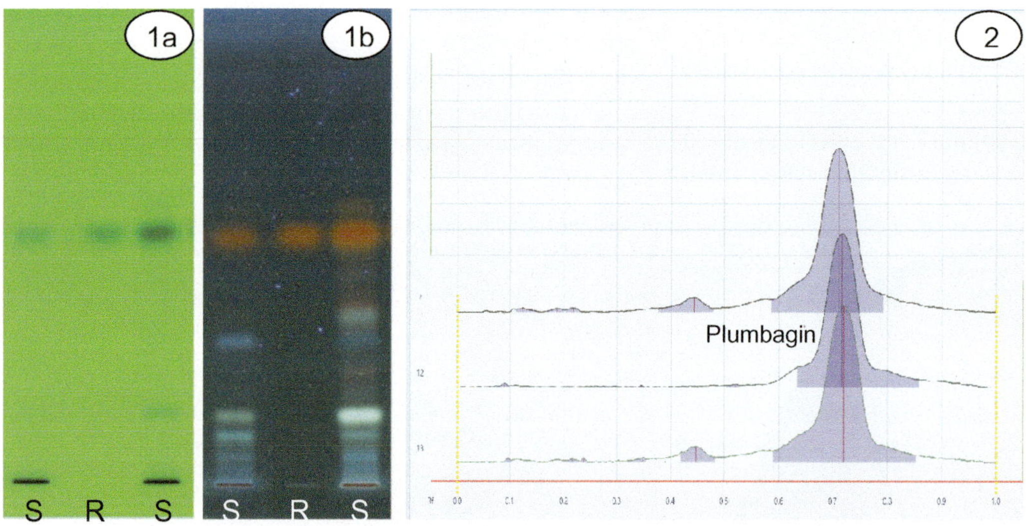

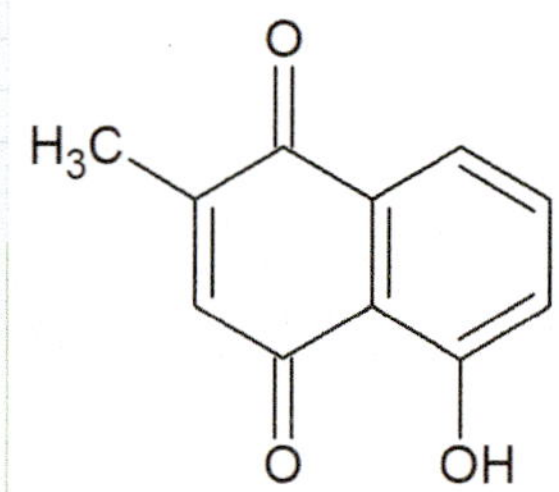

Plumbagin

Plate 1: HPTLC Chromatogram of *Plumbago zeylanica* Sample (S) and Reference Pulmbagin (R).
1a: Plate Visualized under UV254; 1b: Plate Visualized under UV366.

Plate 2: Overlay Spectra of Sample and Standard.

1. Vishnukanta AC (2010). Evaluation of anticonvulasant activity of *Plumbago zeylanica* Linn leaf extract. Asi J Pharma Cli Res 3(1): 76-78.
2. Vijver LM (1971). Antibacterial Activity in Roots of *Plumbago zeylanica*. Planta Med 20: 8-13.
3. Aditi G (1999). Medicinal plants used in traditional medicine in Jimma zone, South West. Ethopia Pharm Biol 37: 321-323.
4. Daiziel JM (1959). The useful plants of west tropical Africa. In: London: The crown agents for the colonies. 424.
5. Jiangsu S (1979). New Medical College, Zhongyao Dictionary (Encyclopedia of Chinese Materia Medica), Scientific and Technological Press, Shanghai. 711-12.
6. Didry N, Dubrevil L, Pinkas M (1994). Activity of anthraquinonic and naphthoquinonic compounds on oral bacteria. Pharmazie 49: 681-3.
7. Gupta A, Gupta A, Singh J (1999). New Naphthoquinones from *Plumbago zeylanica*. Pharma Biol 37: 321-3.
8. Sankaram AVB, Srinivasarao A, Sidhu GS (1976). Chitranone- a new binaphthaquinone from *Plumbago zeylanica*. Phytochem 15: 237-8.

9. Hiradeve S, Danao K, Kharabe V, Mendhe B (2010). Evaluation of anticancer activity of *Plumbago zeylanica* Linn. Leaf extract. IJBR 1(2): 52.
10. Kofi A, Dickson R, Mensah A, Fleischer TC (2009). Acaricidal effect of *Plumbago zeylanica* L. against Amblyoma variegatum. Pharmacog J 1: 190-194.
11. Nile SH (2010). Antioxidant activity and flavanoid derivatives of *Plumbago zeylanica*. J Nat Prod 5: 130-133.
12. Alpana R (1996). Effect of *Plumbago zeylanica* in hyperlipidaemic rabbits and its modification by vitamin E. Indian J Pharmaco 28: 161-166.
13. Bhargava SK (1984). Effects of plumbagin on reproductive function of male dog. Indian J Exp Biol 22: 153-156.

36

Podophyllum hexandrum Royle

Family

Berberidaceae

Vernacular Name

Hindi: Bantrapushi, Giriparpat; English: Himalayan May apple, Indian May apple; Punjabi: Papri, Vankakri.

Distribution [1-2]

Podophyllum hexandrum is native of India, China, Nepal, Bhutan and Afghanistan. In India it is mostly found in the Himalaya from Jammu Kashmir, Uttarakhand, Himanchal Pradesh to Sikkim and Arunachal Pradesh.

Traditional Uses [3]

Traditionally it is used for the treatment of cold, constipation, septic wounds, burning sensation, mental disorders, plague and monocytoid leukemia.

Major Chemical Constituents [4-5]

Quercetin, kaempferol and podophyllotoxin.

Pharmacological Activity [6-8]

Antiviral, antioxidant and antifungal activity.

Plant Description

Podophyllum hexandrum is an erect, glabrous, perennial herb grows up to 40 centimetres. Stem is soft

A Vegetative Plant

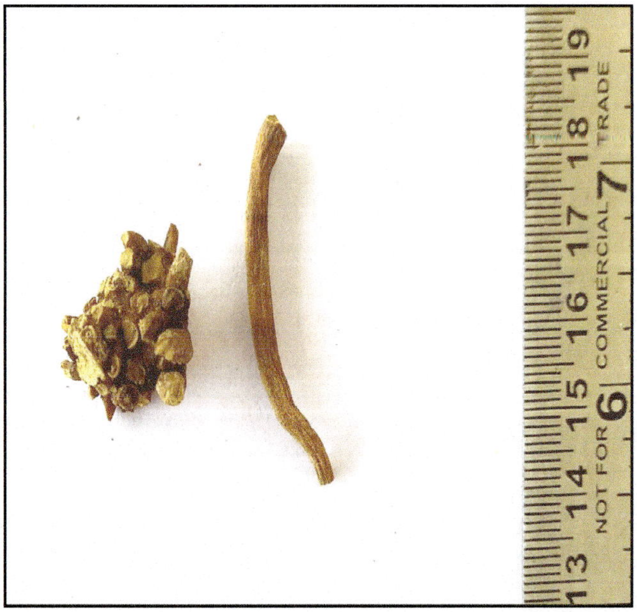

Dried Piece of Root

glabrous, fleshy, succulent, straight and leafy at top. Leaves are big, dark green, orbicular-reniform, palmate, divided in to three lobes. Flowers are solitary, cup shaped, white or pinkish. Rhizome is perennial, yellow coloured, fleshy, irregular and bearing lateral roots.

Anatomy

Transverse section of root is almost circular in outline. The outer layer is made up of irregular cork cells. The cortical region is made up of round or irregular parenchymatous cells occupying major area of the section. Single layer endodermis is clearly seen made of rectangular or ovoidal cells. The xylem is scattered in vascular region, consists of vessels, fibers, tracheids and xylem parenchyma. Vessels are mostly in groups, and surrounded by parenchymatous phloem.

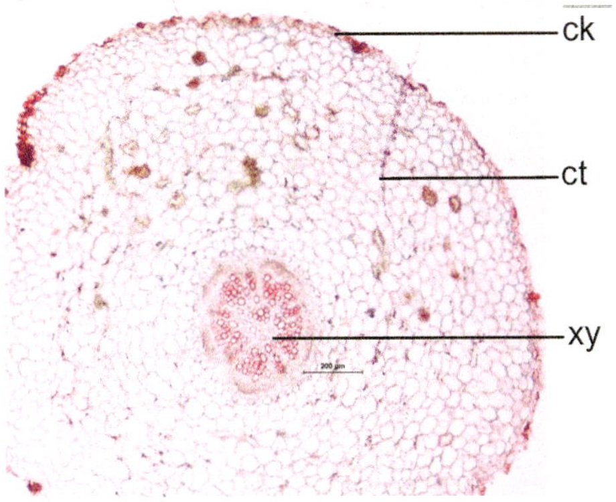

T.S. (4 x)

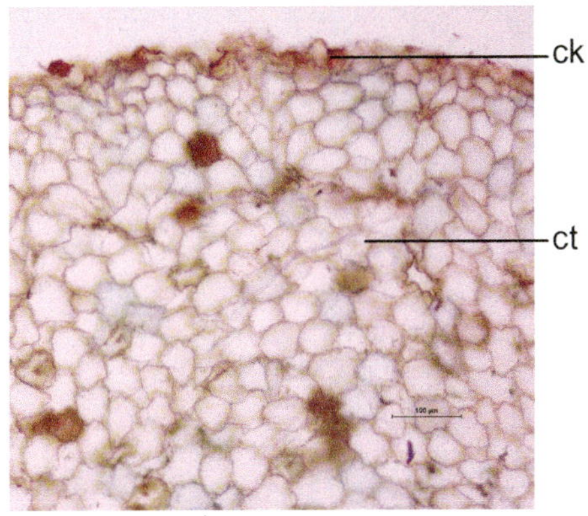

T.S. (10 x)

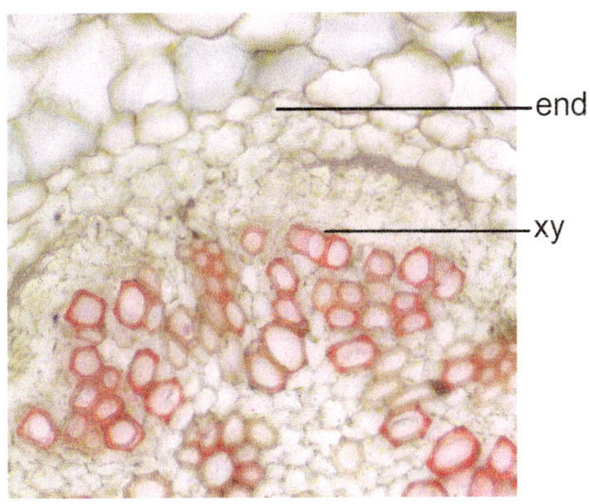

T.S. (20 x)

Transverse Section (T.S.) of *Podophyllum hexandrum* Root.
Abbreviation: ck- cork; ct- cortex; end- endodermis; xy- xylem.

HPTLC Studies

Working Solution

Accurately weighed (5g) Crude powdered sample was subjected to cold extraction with methanol (50 ml), continuously shaken for 6h and then allowed to stand at room temperature (25 ^{0}C ± 2) for 18h. Extraction was repeated thrice, and the pooled filtrate was concentrate in rotatory evaporator (Buchi, Switzerland). Work solution for HPTLC study was made of 10 mg/ml.

Standard Solution

Podophyllotoxin (>98 per cent) of 1 mg/ml strength was freshly prepared

Sample Application

2 µl of standard solution and 10 µl samples were applied as 6 mm band with CAMAG HPTLC applicator on activated silica gel 60 GF 254 HPTLC plate. The chromatogram was developed on pre-saturated CAMAG HPTLC twin trough chamber for 10 min and the solvent system was allowed to run for 8 cm.

Solvent System

Toluene- ethyl acetate- acetic acid (7.5:3.5:0.2 v/v)

Visualization

The chromatogram was visualized at UV_{254} nm and densitometric scanning of podophyllotoxin was done at 290 nm.

Identification of Podophyllotoxin

The standard podophyllotoxin was identified at R_f 0.28.

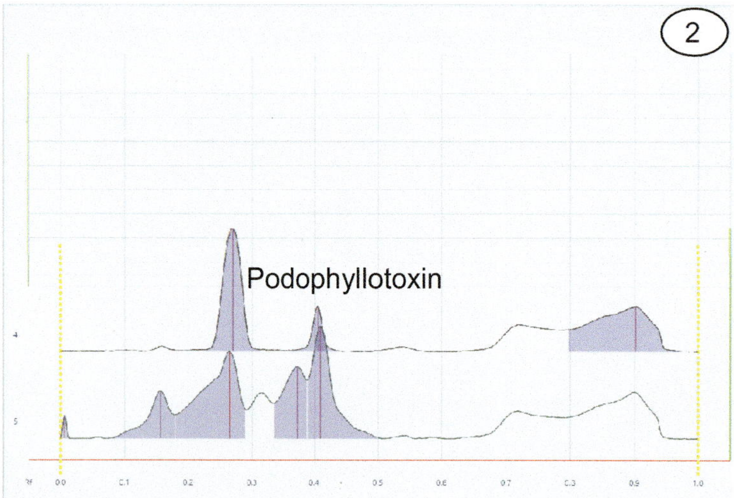

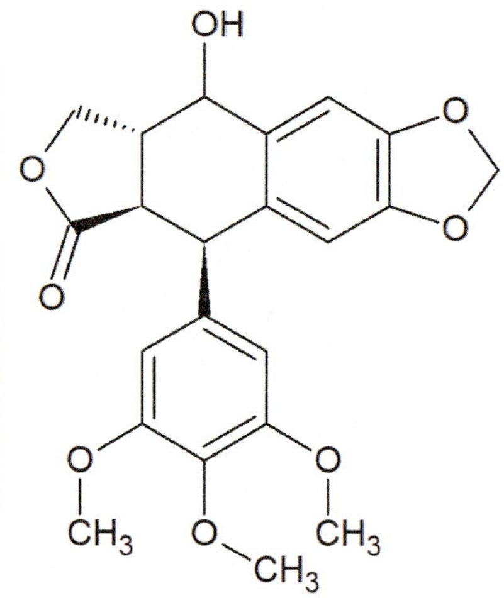

Podophyllotoxin

Plate 1: HPTLC Chromatogram of *Podophyllum hexandrum* (S) and Standard Podophyllotoxin (R)
Plate 2: Overlay Spectra of Sample and Standard.

1. Choudhary DK, Kaul BL, Khan S (1998). Cultivation and conservation of *Podophyllum hexandrum*- An overview. J Med Aro Plant Sci 20:1071-1073.
2. Kumar S, Singh J, Shah NC, Ranjan V (1997). Indian Medicinal Plants Facing Genetic Erosion. Central Institute of Medicinal and Aromatic Plants, Lucknow, CSIR pp. 205.
3. Beutner KR, Vonkrough G (1990). Current status of podophyllotoxin for the treatment of genital warts. Seminars in Dermatology 9: 148.
4. Lerndal T, Svensson B (2000). A clinical study of CPH 82 vs. methotrexate in early rheumatoid arthritis. Rheumatology (Oxford) 39: 316.
5. Imbert TF (1998). Discovery of podophyllotoxins. Biochimi 80: 207-222.
6. Sudo K, Konno K, Shigeta S, Yokota T (1998). Inhibitory effects of podophyllotoxin derivatives on herpes simplex virus replication. Antivir Chem Chemother 9:263-267.
7. Ganie SA, Zargar BA, MasoodA and Zargar MA (2012). Effect of long dose exposure of *Podophyllum hexandrum* methanol extracton antioxidant defense system and body and organ weight changes of albino rats. Asian Pacific J Trop Biomed 2(3): 1600-5.
8. Berkowitz DB, Choi S, Maeng JH (2000). Enzyme-assisted asymmetric total synthesis of (–)-podophyllotoxin and (–)- picropodophylline. J Org Chem 65:847-860.

37

Potentilla fulgens Linn.

Family

Rosaceae

Vernacular Name [1-2]

Hindi: Bajradanti; English: Cinquefoils; Garhwali: Dentamanjari; Meghalaya: Lyniangbru; Nepali: Dentamanjari.

Distribution

In India the species is commonly found in Eastern and Western Himalayas, ranging from Jammu and Kashmir, Uttarakhand, Manipur, Himachal Pradesh, Nagaland and Khasi Hills of Meghalaya up to 1800 meter above from sea level.

Traditional Uses [3]

Root is used for treatment of pyria, colic pain and tumor; aqueous extract is also used for intestinal parasitic infection. In Nepal and Bhutan, plant juice is used for treatment of stomach problems, cough, cold and respiratory problems.

Major Chemical Constituents [4]

Epicatechin, Potentene-A and rutin

Pharmacological Activity [5-7]

Anti-oxidant, Anti-hyperlipidemic and anti-diabetic activity

A Vegetative Plant

Dried Root

Plant Description

Potentilla fulgens L. is a small perennial herb grows mainly in moist places. Leaves are palmate; more than 10 leaflets are arranged pinnately. Stem are cylindrical, soft and, hairy. Flowers are small yellow coloured and fruits are nut. Roots are thick, woody, pale yellow coloured from where secondary roots arise.

Anatomy

Transverse section of mature root is almost circular in outline. The outer layer is made up of cork cells, followed by cortical region are made up of multilayered irregular parenchymatous cells. Oil droplets are observed in cortical region. Just below the cortex 4 to 10 layered thick phloem having sieve tubes, companion cells and phloem fibers are present. The xylem is consists of vessels, fibers, tracheids and xylem parenchyma. Vessels are mostly solitary or in groups of 2 or 5. Medullary rays are thick-walled, radiating, and parenchymatous.

HPTLC Studies

Working Solution

Accurately weighed (5 g) crude powdered sample was subjected to cold extraction with methanol (50 ml), continuously shaken for 6 h and then allowed to stand at room temperature (25 ^{0}C ± 2) for 18 h. The extraction was repeated thrice and the pooled filtrate was concentrate in rotatory evaporator (Buchi, Switzerland). Working solution of 10 mg/ml was freshly prepared on the day of quantative analysis.

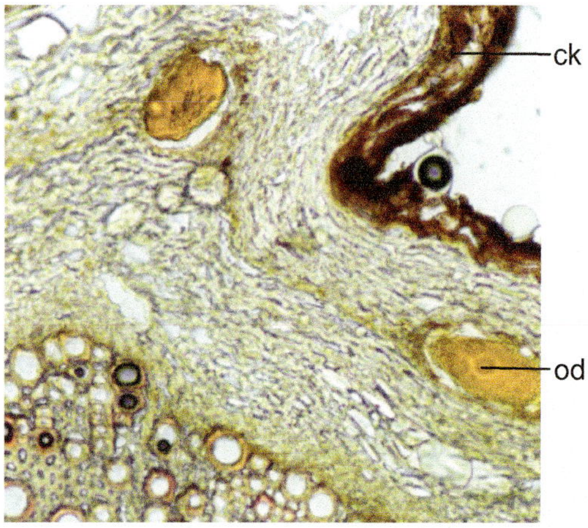

T.S. (20 x)

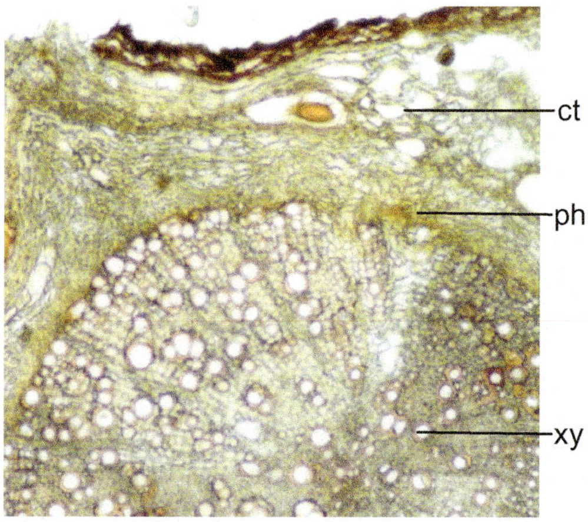

T.S. (4 x)

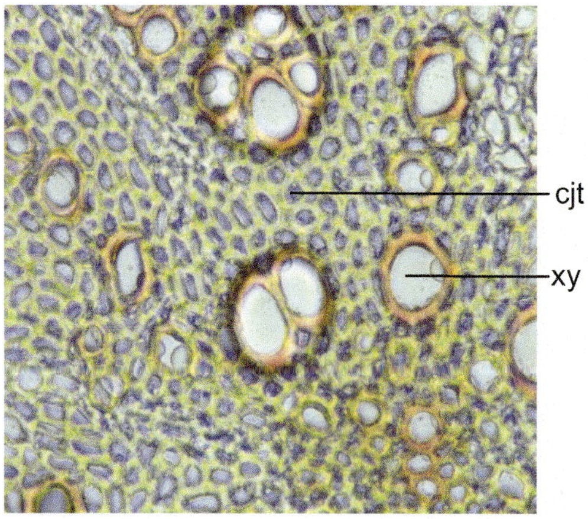

T.S. (20 x)

Transverse Section (T.S.) of *Potentilla fulgens* Root.

Abbreviation:; ck- cork; ct- cortex; cjt- conjunctive tissue; od- oil droplets; ph- phloem; xy- xylem.

Standard Solution

Epicatechin (≥ 98 per cent) was freshly prepared of strength 1 mg/ml.

Sample Application

2 µl of standard solution and 10 µl of sample was applied as 6 mm band with CAMAG HPTLC applicator on activated silica gel 60 GF_{254} HPTLC plate. The chromatogram was developed on pre-saturated CAMAG HPTLC twin trough chamber for 10 min and the solvent system was allowed to run for 8 cm.

Solvent System

Toluene- ethyl acetate- formic acid (5:4:1 v/v)

Visualization

The chromatogram was visualized in UV_{254}nm UV_{366} and densitometric scanning of epicatechin was done at 205 nm.

Identification of Epicatechin

The standard epicatechin was identified at R_f 0.16.

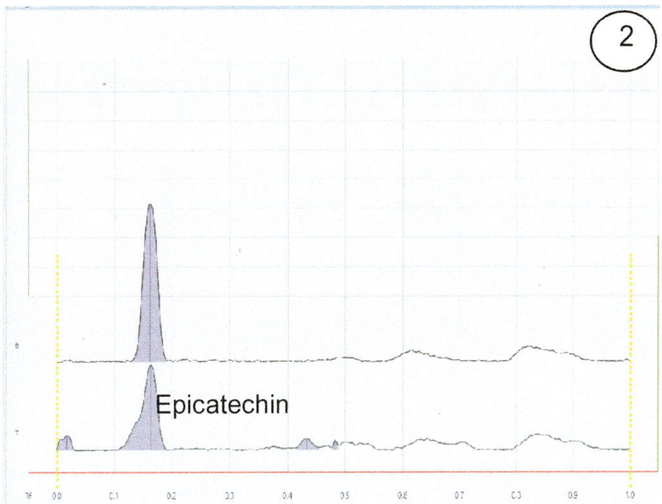

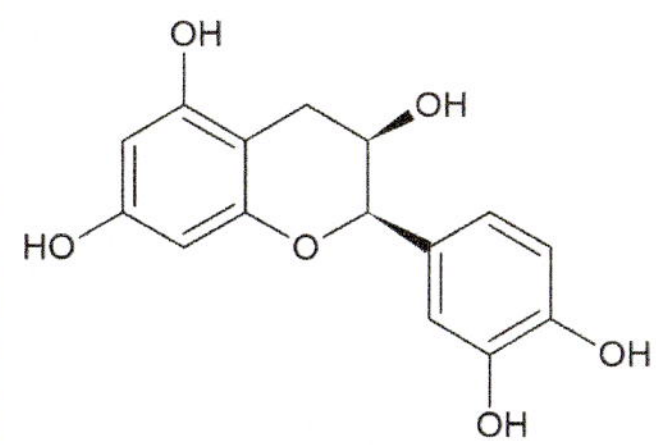

Epicatechin

Plate 1: HPTLC Chromatogram of *Potentilla fulgens* Sample (S) and Standard Epicatechin (R).
1a: Visualized under UV_{254}; 1b: Visualized under UV_{366}.
Plate 2: Overlay Spectra of Sample and Standard.

1. Panigrahi G, Dixit BK (1980). Studies on taxonomy and economic utilization of twelve species of *Potentilla* (Rosaceae) in India, J. Econ Tax Bot 1:127.
2. Manandhar NP (2002). Plants and People of Nepal, Timber Press, Portland, Oregon, USA.
3. Rosangkima G, Rongpi T, Prasad SB (2010). Ethno-medicinal value of some anticancer medicinal plants from north-east India: an in vivo screening in murine tumor model. Sci Vis 10:123-132.
4. Jaitak V, Kaul VK, Himlata, Kumar N, Singh B, Dhar J, Sharma OP (2010). New hopane triterpenes and antioxidant constituents from *Potentilla fulgens*, Nat Prod Comm 5(10): 1561-1566.
5. Syiem D, Syngai G, Khup PZ (2002). Hypoglycemic effects of *Potentilla fulgens* in normal and alloxan-induced diabetic mice. J Ethnophar 83:55-61.
6. Jaitak V, Sharma K, Kalia K, Kumar N, Singh HP, Kaul VK, Singh B (2010). Antioxidant activity of *Potentilla fulgens* : An alpine plant of western Himalaya, J Food Compos Anal 23(2): 142.
7. Syiem D, Khup PZ, Syiem AB (2009). Effects of *Potentilla fulgens* Linn. On carbohydrate and lipid profiles in diabetic mice, Pharmacologyonline 2:787-795.

38

Pueraria tuberosa (Roxb. Ex Willd.) DC.

Family

Fabaceae

Vernacular names

Sanskrit: Gandhaphala, Bhukushmandi; Hindi: Vidarikand, Vendari, Bhadra; Bengali: Bedarikand, Patalkand; Marathi: Virdari; English: Indian Kudzu; Telugu: Darigummadi.

Distribution [1-2]

Pueraria tuberosa (Roxb. Ex Willd.) DC. is native of India, Nepal and Pakistan. It is distributed throughout the Indian Himalayas ranging from Sikkim and, Kumaon regions to the tropical and subtropical region, up to an elevation of 4000 ft.

Tuber in Natural Habitat

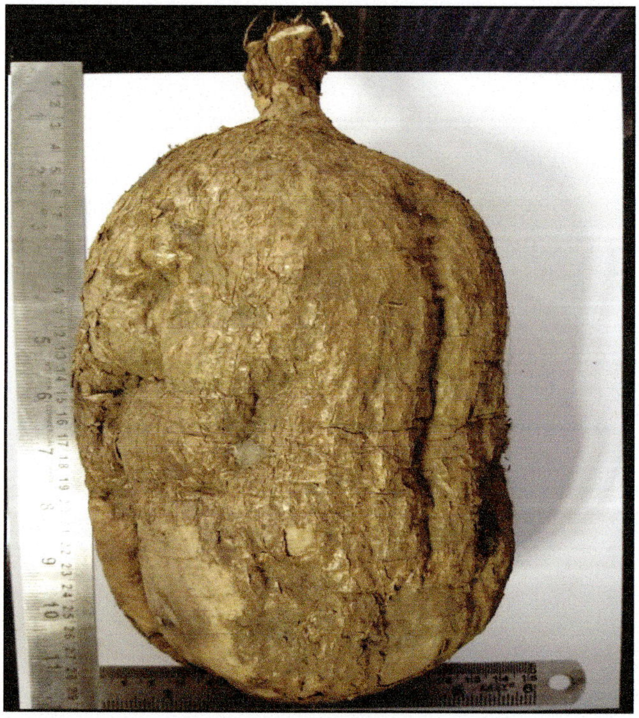

Fresh Tuber

Traditional Uses [3]

In Ayurveda the tuber is used in burning sensation, spermatorrhoea, intrinsic hemorrhage, tuberculosis and heart diseases; it also used in sexual weakness, inflammation, fever, diarrhoea, chest pain, rheumatism and skin diseases.

Major Chemical Constituents [4-5]

Puerarin, daidzein, genistein, stigmasterol and tuberosin

Pharmacological Activity [6-9]

Immunomodulatory, anti-fertility, antioxidant and anti-diabetic activity

Morphological Description

Pueraria tuberosa is an herbaceous climber, succulent perennial herb, woody stem with tuberous root with very large tuber. Leaves are large trifoliate, alternate, oval shaped leaflet, smooth at upper surface. Flowers are small, blue or purple bluish in colour and fruit pod is linear. The tubers are big pot like, globose; outer surface is very hard, brown colored, thick and inner surface is soft, fibrous, white colored, starchy and sweet taste.

Anatomy

Transverse section of tuber is almost circular in outline, shows very narrow range of cork and wider cortex, occupying major area of section, well distinguished endodermal layer and wide stellar system. Detailed structure accounted that epidermis is not discernible followed by multi-layered thin-walled cork cambium lie underneath this. Well developed endodermis and fibrous pericycle are present beneath the multilayer's stone cell filled with calcium oxalate crystals. Phloem is narrow and well developed consists sieve tube, companion cells, fibers and parenchyma. Xylem consists of vessels, tracheids, fibers and parenchyma. Medullary rays are broad, multiseriate and parenchymatous, filled with starch grains which is oval or polygonal in shape.

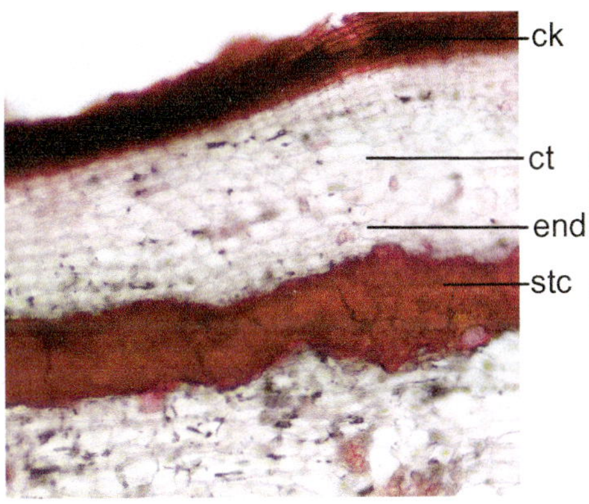

T.S. (10 x)

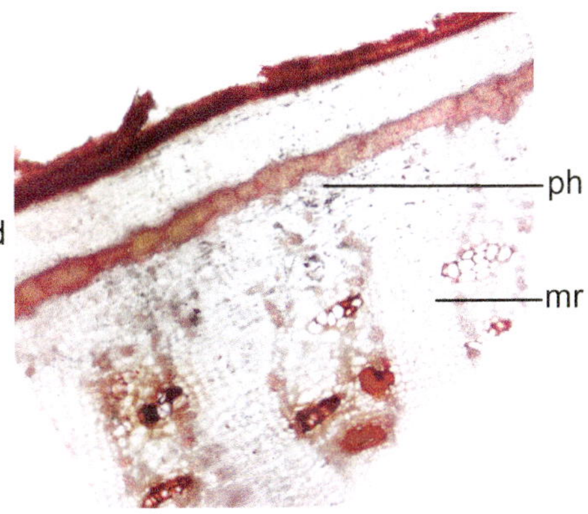

T.S. (4 x)

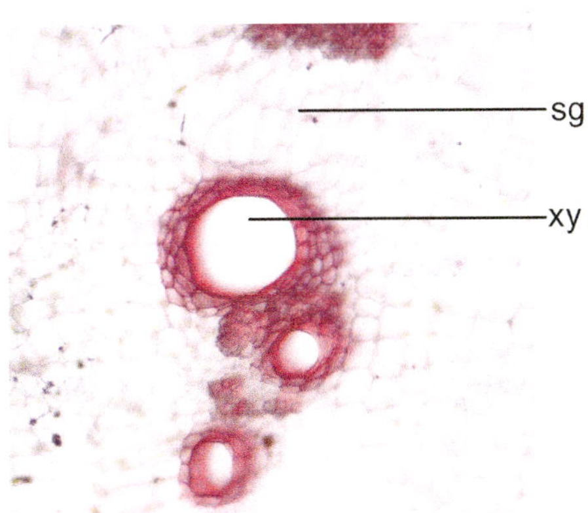

T.S. (40 x)

Transverse Section (T.S.) of *Puraria tuberosa* Tuber

Abbreviation: ck- cork; ct- cortex; end- endodermis; mr- medullary rays; ph- phloem; sg- starch grains; stc- stone cells; xy- xylem.

HPTLC Studies

Sample Solution

Accurately weight (5 g) crude powdered sample was extracted (cold) with methanol (50 ml) followed by continuously shaken for 6 h and then allowed to stand for 18 h at room temperature (25 °C ± 2). The extraction was repeated thrice, filtered and the pooled filtrate was concentrate in rotatory evaporator (Buchi, Switzerland), dried residue was stored at 4 °C. Working solution of 10 mg/ml was freshly prepared on the day of quantative analysis.

Standard Solution

Peurarin (≥ 98 per cent) and diadzein (≥ 98 per cent) were freshly prepared of strength 1 mg/ml.

Sample Application

2 µl of standard solution and 10 µl of sample was applied as 6 mm band with CAMAG HPTLC applicator on activated silica gel 60 GF_{254} HPTLC plate. The chromatogram was developed on pre-saturated CAMAG HPTLC twin trough chamber for 15 min and the solvent system was allowed to run for 8 cm.

Solvent System

Chloroform: methanol (8:2 v/v)

Visualization

The chromatogram was visualized at UV_{366} and densitometric scanning of Peurarin and diadzein was done at 250 nm.

Identification of Puerarin and Diadzein

The standards puerarin and diadzein was identified at R_f 0.12 and 0.62 respectively.

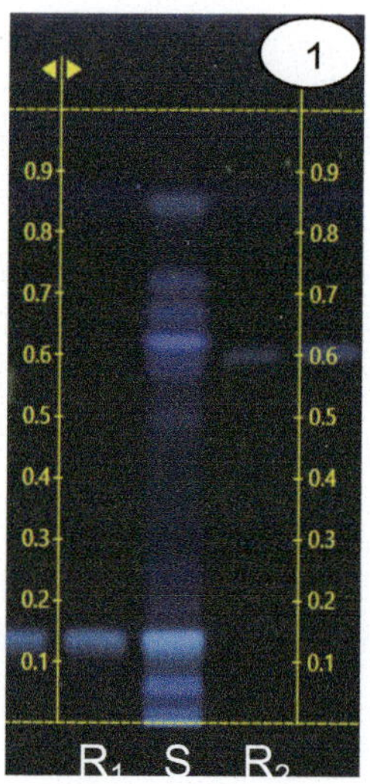

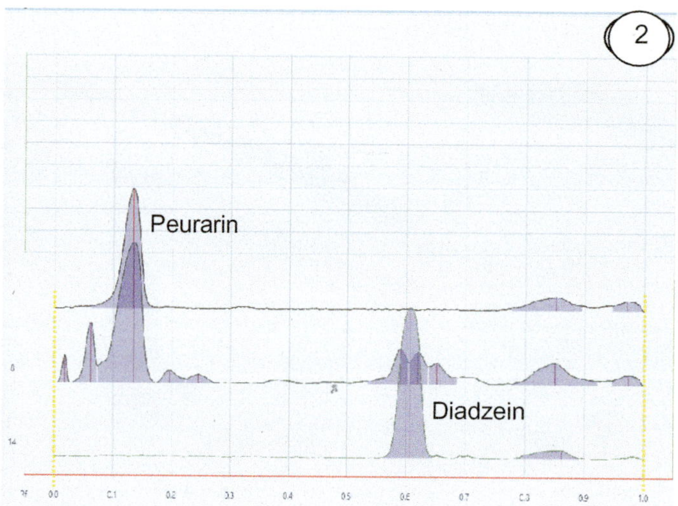

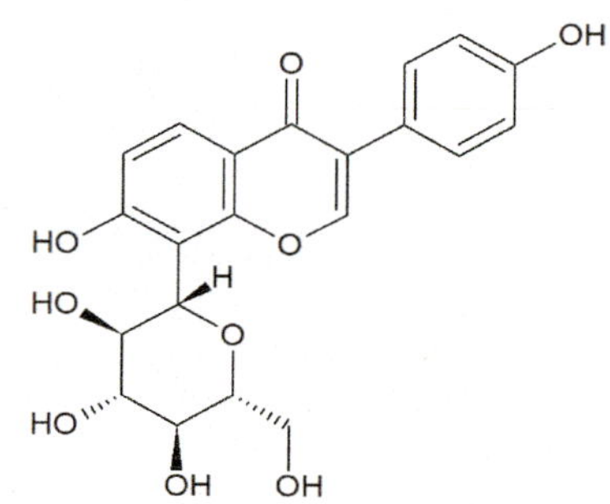

Peurarin

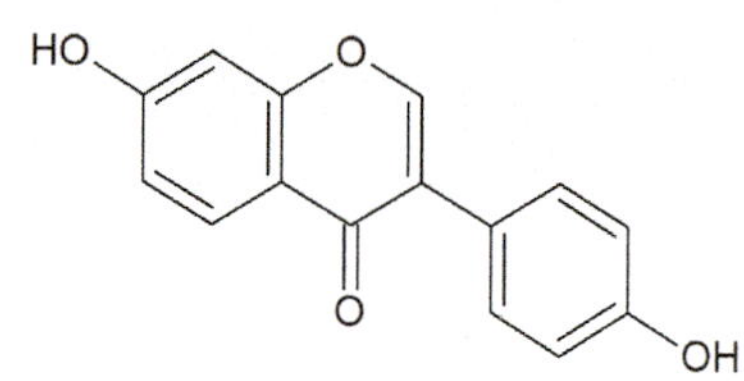

Diadzein

Plate 1: HPTLC Chromatogram of *Pueraria tuberosa* Sample (S) and Standard Puerarin (R1) and Diadzein (R2). Plate 2: Overlay Spectra of Sample and Standard.

1. The Ayurvedic Pharmacopoeia of India, Part- I Volume – II and V, Government of India Ministry of Health And Family Welfare: 173 and 224.
2. Keung WM (2002). *Pueraria*: the genus *Pueraria*. 1st ed. London: Taylor and Francis.
3. Rao NV, Pujar B, Nimbal SK, Shantakumar SM, Satyanarayana S (2008). Nootropic activity of tubar extract of *Pureraria tuberosa* (Roxb). Indian J Exp Biol 46: 491-8.
4. Maji AK, Pandit S, Banerji P, Banerjee D (2014). *Pueraria tuberosa*: a review on its phytochemical and therapeutic potential. Nat Prod Res 28:2111-27.
5. Khan R, Agrawal P, Kapil R (1996). Puetuberosanol, an epoxychalcanol from *Pueraria tuberosa*. Phytochem 42: 243-4.
6. Maji AK, Mahapatra S, Banerjee, D (2014). In-Vivo Immunomodulatory Potential of Standardized *Pueraria tuberosa* Extract And Its Isoflavonoids. Int J Phar Pharmace Sci 6: 861-67.

7. Tripathi AK, Kohli, S (2013). Anti-Diabetic Activity and Phytochemical Screening of Crude Extracts of *Pueraria tuberosa* DC. (FABACEAE) Grown in India on STZ -Induced Diabetic Rats. Asi J pharmaceu Med res 3: 13, 66-73.

8. Pandey N, Chaurasia JK, Tiwari OP, Tripathi YB (2007). Antioxidant properties of different fractions of tubers from *Pueraria tuberosa* Linn. Food Chem 105: 219–222.

9. Prakash AO, Saxena V, Shukla S, Mathur R (1985). Contraceptive potency of *Pueraria tuberosa* D.C. and its hormonal status. Acta Eur Fertil 16: 59–65.

39

Rauvolfia serpentina L. Benth. Ex Kurz.

Family

Apocynaceae

Vernacular name

Sanskrit: Sarpaghandha; Hindi: Chandrabhaga, Sarpaghandha; Bengali: Chandra; English: Indian snake root; Tamil: chevanamalpodi.

Distribution [1]

Rauwolfia serpentina is distributed in the foot -hill of Himalaya range, up to the elevation of 1300–1400 m.

Traditional Uses [1-3]

Sarpagandha is widely used in the medical system and is also used in Ayurveda, Unani and folk medicine in treatment of poisonous snakes bite. It is also used in general treatment for high blood pressure, insomnia and is also used on various central nervous system disorders, both physical and motor, including anxiety and excitement. Root extracts are used to treat intestinal disorders, such as diarrhea and dysentery.

Flowering Twig

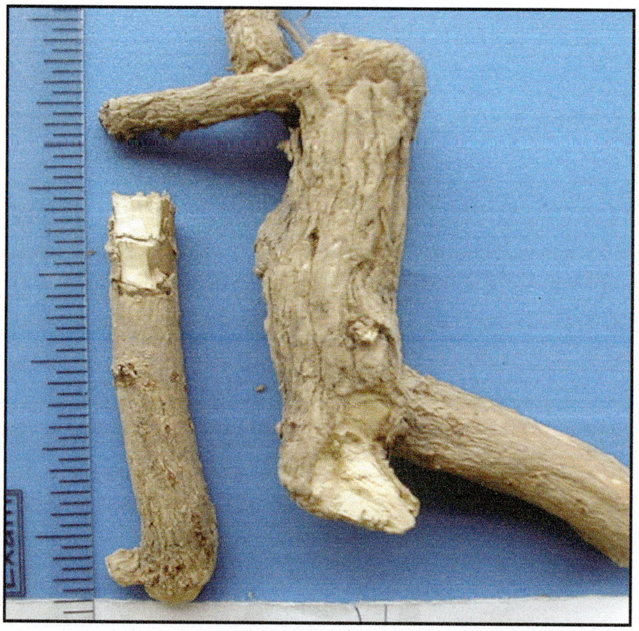

Dried Root

Major Chemical Constituents [4-5]

Reserpine, ajmaline, serpentine and rauwolfinine.

Pharmacological Activity [6-8]

Anti-hypertensive, antipsychotic and anticancer activity.

Plant Description

Rauvolfia serpentina L. Benth. Ex Kurz. is an erect perennial shrub with a long, irregularly, nodular, yellowish root stock. Twigs are erect cylindrical, smooth and covered with friable bark. Leaves are lanceolate, acute bright green above and pale beneath and present in whorl of 3. Flowers are numerous; bracts small, irregular corymbose cymes, white, often tinged with violet. Fruit is small drupe, single or didymous, shining black. The roots are thick, woody, cylindrical, covered with a thin friable bark. Bark is internally smooth, rough, odorless and bitter in taste.

Anatomy

Transverse section of mature root is circular in outline. The outer layer is made up of rectangular 20 or more layered cork cells. The cortical region is made up of 25 to 28 layered irregular parenchymatous cells; cortical cells are broad having intercellular spaces. Just below the cortex 6 to 10 layered thick phloem having sieve tubes, companion cells and phloem fibers. Vascular cambium is 2 or 3 layered. The xylem is 2 or more celled wide and consists of vessels, fibers, tracheids and xylem parenchyma. The medullary rays are thick-walled, radiating, and parenchymatous containing starch grains. In longitudinal section vessels are seen vertically running broad segmented canals with bordered pitted and thickened wall, parenchyma are seen as narrow elongated tubes.

HPTLC Studies

Sample Solution

Accurately weight (5 g) crude powdered sample was subjected to cold extraction with methanol (50 ml), continuously shaken for 6 h and then allowed to stand at room temperature ($25^0C \pm 2$) for 18 h. The extraction was repeated thrice and the pooled filtrate was concentrate in rotatory evaporator (Buchi, Switzerland). Working solution of strength 10 mg/ml was freshly prepared on the day of quantative analysis.

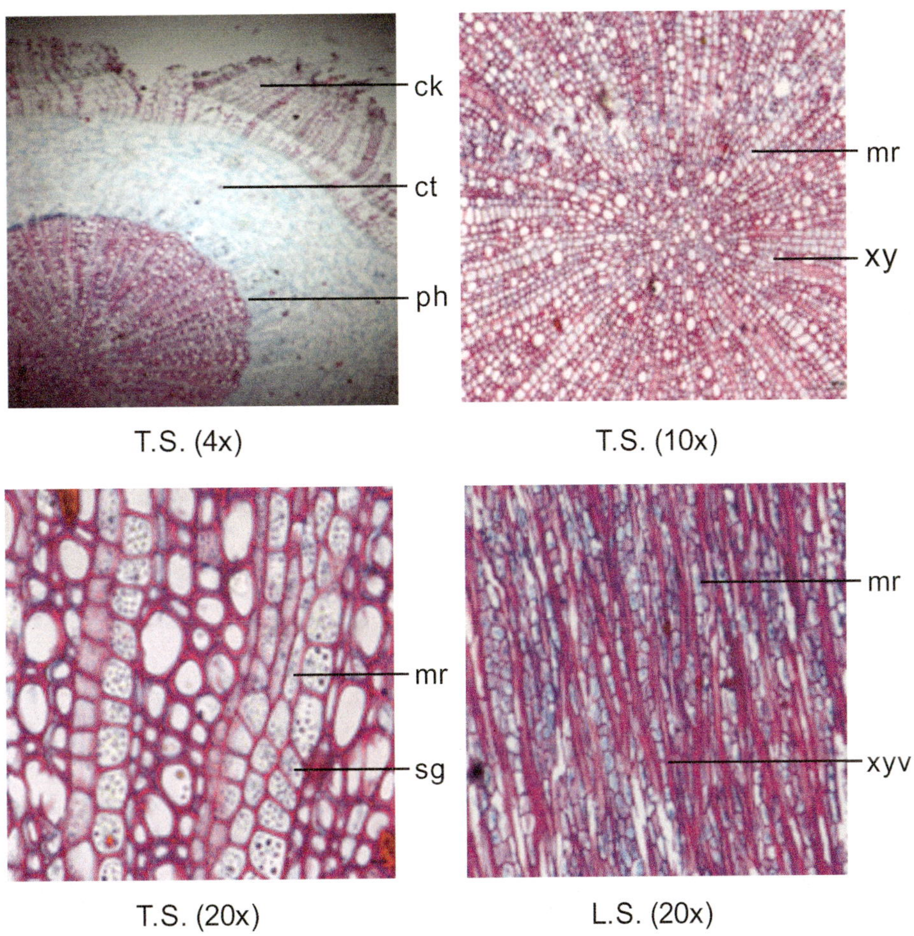

Transverse Section (T.S.) & Longitudinal Section (L.S.) of *Rauvolfia serpentina* Root.

Abbreviation: ck- cork; ct- cortex; mr- medullary rays; ph- phloem; sg- starch grains; xy- xylem; xyf-xylem fibre.

Standard Solution

Reserpine (≥ 98 per cent) was freshly prepared of strength 1 mg/ml.

Sample Application

2 µl of standard solution and 10 µl of sample was applied as 6 mm band with CAMAG HPTLC applicator on activated silica gel 60 GF_{254} HPTLC plate. The chromatogram was developed on pre-saturated CAMAG HPTLC twin trough chamber for 20 min and the solvent system was allowed to run for 8 cm.

Solvent System

Toluene: ethyl acetate: diethyl amine (7:2:1 v/v)

Visualization

The chromatogram was visualized at UV_{366} and densitometric scanning of reserpine was done at 270 nm.

Identification of Reserpine

The standard reserpine was identified at Rf 0.5.

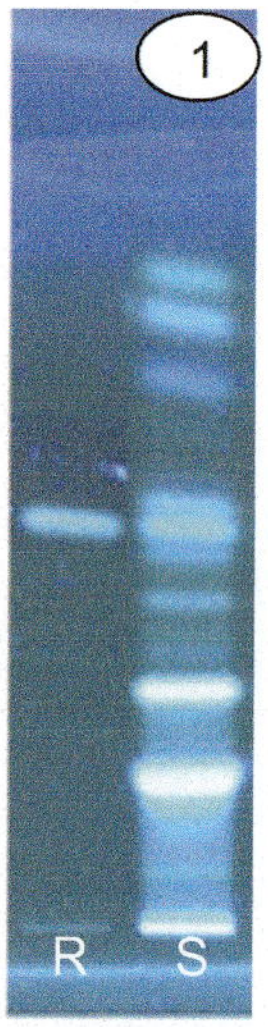

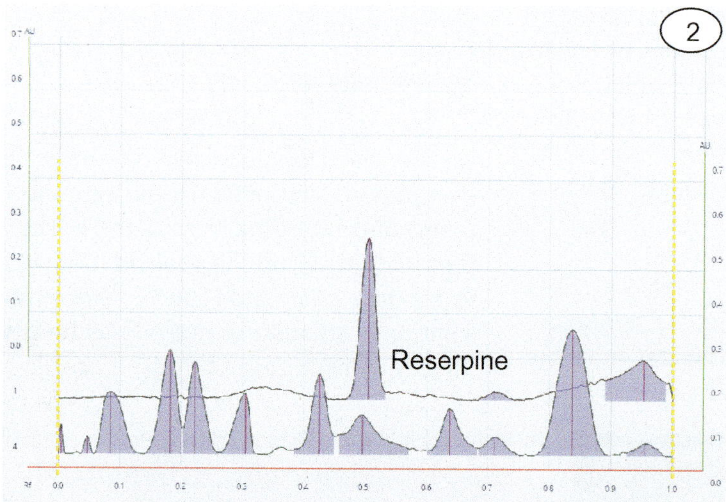

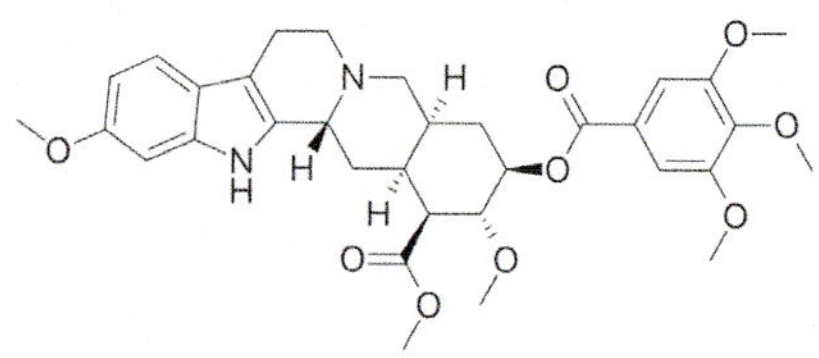

Reserpine

Plate 1: HPTLC Chromatogram of *Rauvolfia serpentina* Sample (S) and Standard Reserpine (R).
Plate 2: Overlay Spectra of Sample and Standard.

1. Biradar N, Hazar I, Chandy V (2016). Current Insight to the Uses of *Rauwolfia*: A Review. Research and Reviews: A J Pharmacog 3(3).
2. Integrative Medicine: A clinicians Journal Douglas lobay, Bsc, ND, Naturopathic physician.
3. Vakil RJ (1955). *Rauwolfia serpentina* in the treatment of high blood pressure: a review of the literature. Circulation 12(2): 220-9.
4. Pandey VP, Cherian E, Patani G (2010). Effect of growth regulators and culture conditions on direct root induction of *Rauvolfia serpentina* L. (Apocynaceae) Benth. by leaf explants, Tro J Pharma Res 9(1): 27-34.
5. Prusoff WH (1961). Effect of reserpine on the 5-hydroxytryptamine and adenosinetriphosphate of the dog intestinal mucosa. Brit J Pharmaco 17: 87-91.
6. Sen G, Bose K (1931). *Rauwolfia Serpentina*, a new. Indian drug for insanity and Hypertension. Indian M World 21: 194.
7. Poonam Agrawal S, Mishra S (2013). Physiological, biochemical and modern biotechnological approach to improvement of *Rauvolfia serpentina*. J Pharm Biol Sci 6(2): 73–78.
8. Kirtikar KR, Basu BD (1993). Indian Medicinal Plants, 2 Ed, Dehra Dun Publishers, Calcutta, India 289.

40

Ricinus communis Linn.

Family

Euphorbiaceae

Vernacular Name [1]

Hindi: Arand, Erand, Andi; Marathi: Errandi; Bengali: Verenda; English: Castor plant, Palm of Christ; Guajarati: Diveli; Punjabi: Arind; Urdu: Be danjir, Arand.

Distribution [2]

Ricinus communis is native of Africa and cultivated in many countries. It is found throughout India in cultivated as well as wild up to 2400 meter.

Traditional Uses [3]

Ricinus communis is used by different communities for the treatment of hypoglycemia, rheumatism, headache, asthma, dermatitis, ringworm, warts and dandruff.

Major Chemical Constituents [4-5]

Lupeol, gallic acid, epicatechin, indole-3-acetic acid and ellagic acid

Pharmacological Activity [6-8]

Antiasthmatic, antidiabetic, antioxidant and anti-inflammatory activity

Plant in Natural Habitat

Dried Piece of Root

Morphological Description

Ricinus communis is fast growing, soft wooded perennial shrub grows up to 5 meter. Young leaves are reddish and old leaves are green in colour contain 5-12 deep lobes, alternate and palmate. Stem is round, soft, straight and hollow. Inflorescence is spike and flowers are monoecious. The fruits are three celled thorny capsule, covered with soft spine like structure. The root is well developed tap root; smooth, whitish and has a hard texture and bitter taste.

Anatomy

Transverse section of root is circular in outline, shows very narrow range of cork and central wood, encircled by narrow phloem, wider cortex. Detailed structure accounted that the root shows narrow cork made of 2 to 5 rows of rectangular cells, followed by wide parenchymatous cortex filled with starch grains. Phloem is narrow and parenchymatous. The xylem is polyarch, with 2-4 cells wide. Lignified medullary rays are present in continuation with phloem, xylem vessels, thin walled fibres and parenchyma. In longitudinal section, medullary rays are heterogeneous and arrangements horizontally, along with vessels and fibers; tracheids are found along with the xylem vessels.

HPTLC Studies

Working Solution

Accurately weigh (5 g) crude powdered was extracted (cold) with methanol (50 ml) followed by continuously shaken for 6 h and then allowed to stand for 18 h at room temperature (25 °C ± 2). The extraction was repeated thrice and the pooled filtrate was concentrate in rotatory evaporator (Buchi, Switzerland), dried residue was stored at 4 °C. Working solution of 10 mg/ml was freshly prepared on the day of quantative analysis.

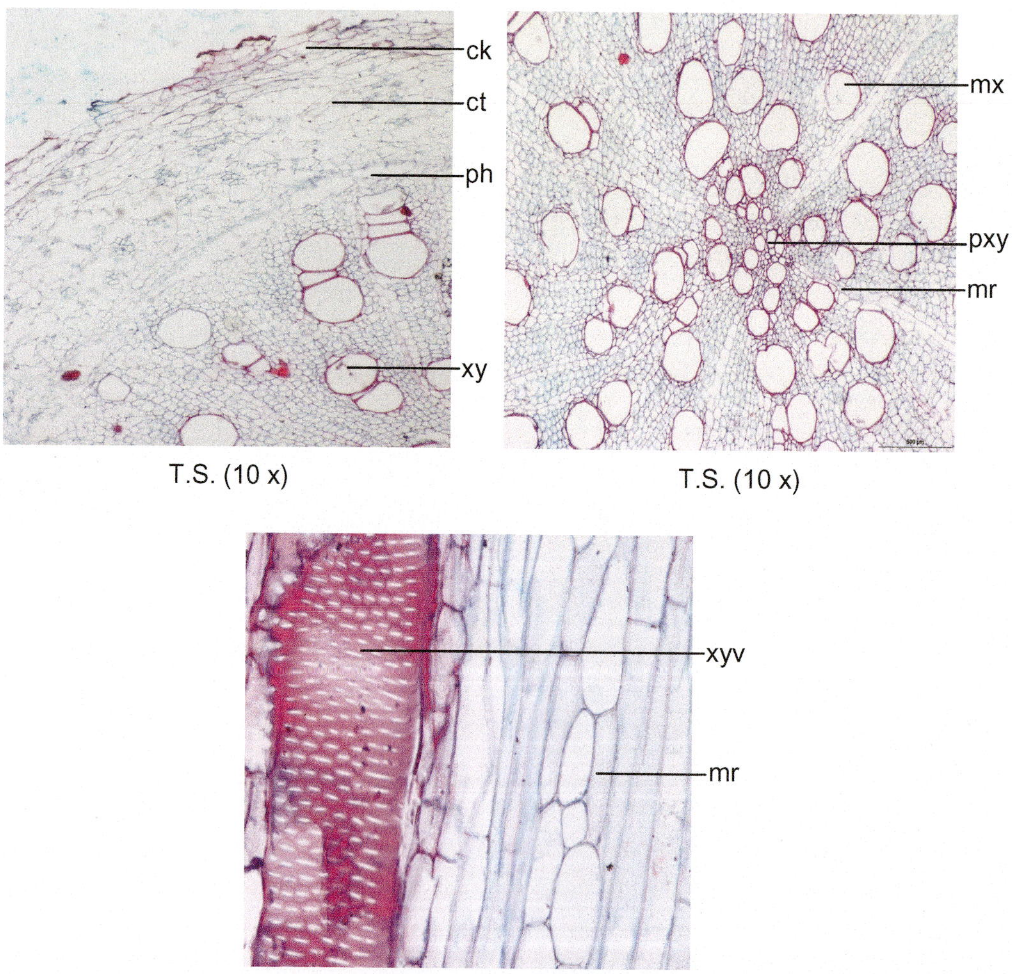

Transverse Section (T.S.) & Longitudinal Section (L.S.) of *Ricinus communis* Root.
Abbreviation: ck- cork; ct- cortex; mr- medullary rays; mx- meta xylem; ph- phloem; pxy- protoxylem; xy- xylem; xyv- xylem vessels.

Standard Solution

Lupeol (≥ 98 per cent) was freshly prepared of strength 1 mg/ml.

Sample Application

2 μl of standard solution and 10 μl of sample was applied as 6 mm band with CAMAG HPTLC applicator on activated silica gel 60 GF_{254} HPTLC plate. The chromatogram was developed on pre-saturated CAMAG HPTLC twin trough chamber for 10 min and the solvent system was allowed to run for 8 cm.

Solvent System

Toluene- ethyl acetate- formic acid (7:3:0.5 v/v).

Visualization

The chromatogram was visualized after derivatization with anisaldehyde sulphuric acid reagent and densitometric scanning was done at 530 nm.

Identification of Lupeol

The standard lupeol was identified at R_f 0.82.

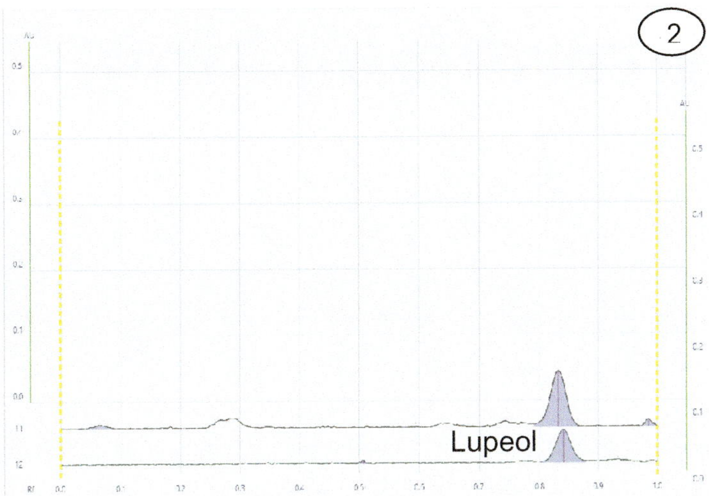

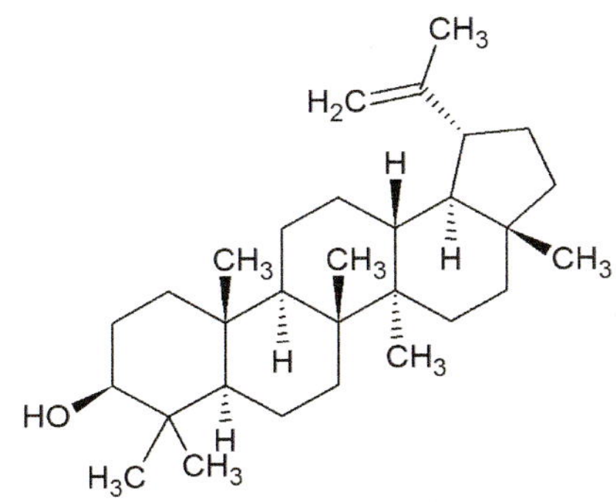

Lupeol

Plate 1: HPTLC Chromatogram of *Ricinus communis* (S) and Standard Lupeol (R)
Plate 2: Overlay Spectra of Sample and Standard.

1. Lal R, Harini A (2017). The Castor Plant- A Review. Int J Ayurvedic Herbal Med 7(1): 2449-2452.
2. Ahmad N, Mishra A, Ahsan, F, Mahmood T, Hasan N, Khan Z (2016). *Ricinus communis*: Pharmacological actions and marketed medicinal products. World J Pharmaceutical Life Sciences 2(6): 179-188.
3. Rana M, Dhamija H, Prashar B, Sharma S (2012). *Ricinus communis* L.- A Review. Int J Pharm Tech Res 4 (4):1706-1710.
4. Singh PP, Ambika Chauhan SMS (2009). Activity guided isolation of antioxidants from the leaves of *Ricinus communis* L. Food Chem 114(3): 1069-1072.
5. Srivastava P, Jyotshna, Gupta N, Maurya AK, Shanker K. New anti-inflammatory triterpene from the root of Ricinus communis. Natural product research. 2014 Mar 4;28(5):306-11.
6. Taur DJ, Patil RY (2011). Antiasthmatic activity of *Ricinus communis* L. roots. Asian Pacific j tropical biomed 1(1):S13-6.
7. Shokeen P, Anand P, Murali YK and Tandon V (2008). Antidiabetic activity of 50 per cent ethanolic extract of *Ricinus communis* and its purified fractions. Food and Chemical Toxicology 46: 3458-3466.
8. Ilavarasan R, Mallika M and Venkataraman S (2006). Anti-inflammatory and free radical scavenging activity of *Ricinus communis* root extract. Journal of Ethnopharmacology 103: 478-480.

41

Rubia cordifolia Linn.

Family

Rubiaceae

Vernacular name

Sanskrit: Manjistha; Hindi: Majith; Bengali: Majith; English: Indian madder; Tamil: Manditti.

Distribution [1]

Rubia cordifolia Linn. is distributed throughout the plains of North India and Western Ghats, up to an altitude of 3500 meter.

Traditional Uses [2–3]

Traditionally, the roots are used to treat various ailments such as bronchitis, rheumatism, kidney stone, dysentery, diabetes, skin problem and blood related disorders.

Major Chemical Constituents [4–5]

Oleanolic acid, purpurin, manjistin, garancin and alizarin.

Pharmacological Activity [6–8]

Anti-oxidant, anti-acne, anti-inflammatory and hepatoprotective activity.

Plant Description

Rubia cordifolia is a perennial, soft prickly stem, climber up to 6 meter. Leaves are glabrous, cordate-ovate to ovate-lanceolate, petiolate and arranged in a whorl of

A Leafy Twig

Dried Root

4-6. Stem is quadrangular, glabrous or hairy, soft and long. Inflorescence is axillary, arises from stem node contains minute white or greenish yellow flower. Root is long cylindrical, smooth and reddish.

Anatomy

Transverse section of mature root is circular in outline. The outer layer is made up of rectangular 4 to 6 layered cork cells. The cortical region is broad and made up of 12 to 18 layered irregular parenchymatous cells, contains round or oval starch grains. Just below the cortex 4 to 8 layered thick phloem having sieve tubes, companion cells and phloem fibers are present. The xylem is 8-10 celled wide and consists of vessels, fibers, tracheids and xylem parenchyma. Vessels are mostly solitary or in groups of 2 or 3 and distributed uniformly. The medullary rays are thick-walled, radiating, and parenchymatous. In longitudinal section vessels are seen vertically running broad segmented canals with bordered pitted and thickened wall, parenchyma is seen as narrow elongated tubes.

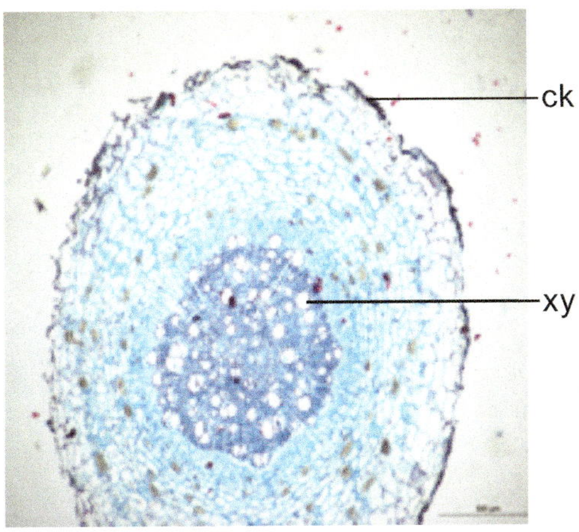

T.S. (4 x)

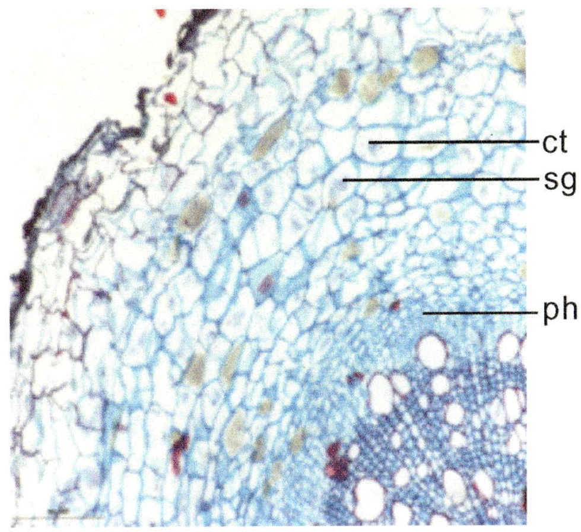

T.S. (10 x)

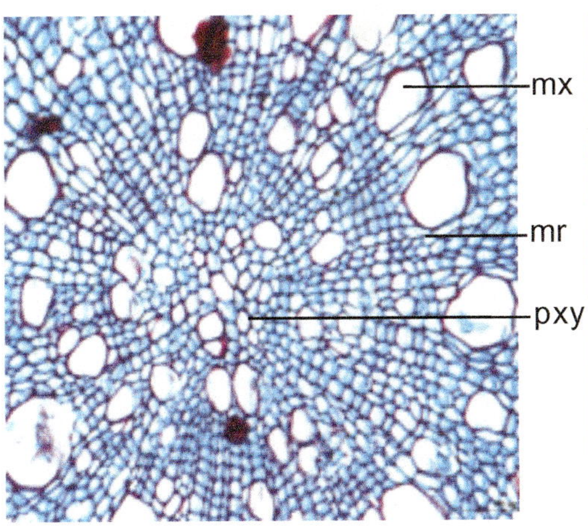

T.S. (20 x)

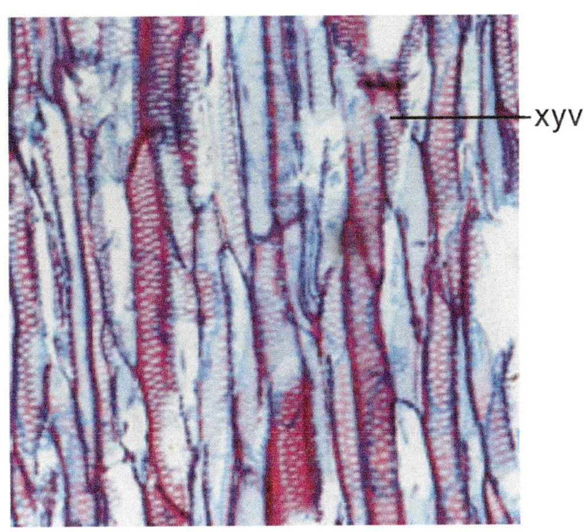

L.S. (20 x)

Transverse Section (T.S.) & Longitudinal Section (L.S.) of *Rubia cordifolia* Root.

Abbreviation: ck- cork; ct- cortex; mr- medullary rays; mx- meta xylem; ph- phloem; pxy- proto xylem; sg- starch grains; xy- xylem; xyv- xylem vessels.

HPTLC Studies

Sample Solution

Accurately weight (5 g) crude powdered sample was subjected to cold extraction with methanol (50 ml) continuously shaken for 6 hrs and then kept at room temperature for 18 hrs. The extraction process was repeated thrice, filtered and the pooled filtrate was concentrate in rotatory evaporator (Buchi, Switzerland). Working solution of strength 10 mg/ml was freshly prepared on the day of quantative analysis.

Standard Solution

Oleanolic acid (≥ 98 per cent) was freshly prepared of strength 1 mg/ml.

Sample Application

2 μl of standard solution and 10 μl of sample was applied as 6 mm band with CAMAG HPTLC applicator on activated silica gel 60 GF_{254} HPTLC plate. The chromatogram was developed on pre-saturated CAMAG HPTLC twin trough chamber for 15 min and the solvent system was allowed to run for 8 cm.

Solvent System

Toluene- ethyl acetate- methanol- formic acid (7:2:0.5:0.5 v/v).

Visualization

The chromatogram was visualized after derivatization with anisaldehyde sulphuric acid reagent and densitometric scanning of oleanolic acid was done at 530 nm.

Identification of Oleanolic Acid

The standard oleanolic acid was identified at R_f 0.35.

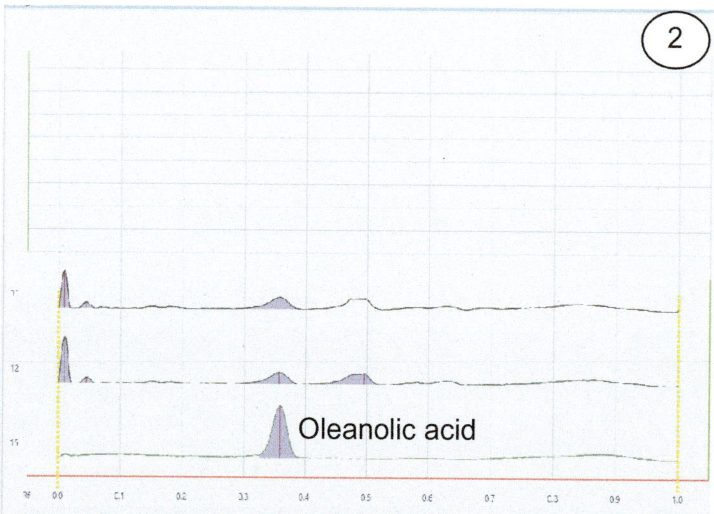

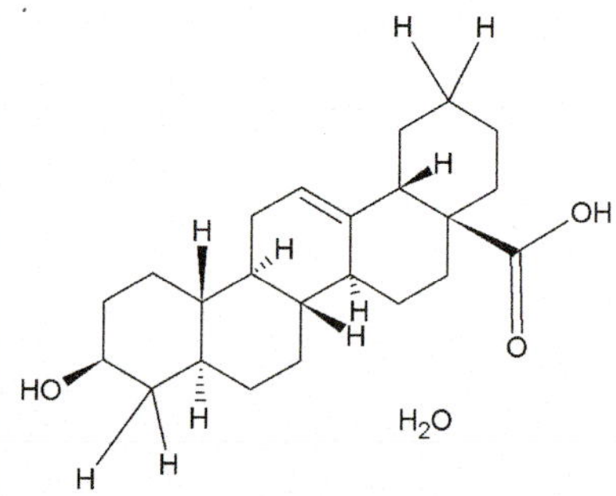

Oleanolic acid

Plate 1: HPTLC Chromatogram of *Rubia cordifolia* Sample (S) and Standard Oleanolic Acid (R).
Plate 2: Overlay Spectra of Sample and Standard.

1. Kirtikar KR, Basu BD (1980). Indian Medicinal Plants. Vol II. 2nd edition. International Book Distributors, Dehradun 1305-1307.
2. Dev S (2006). A selection of Prime Ayurvedic plant drugs Ancient-Modern Concordance, New Delhi: Anamaya publishers.
3. Sivarajan VV, Balachandran I (1994). Ayurvedic Drugs and Their Plant Sources, Oxford and IBH Publishing Co.
4. Anonymous (1999). The Wealth of India – Raw Materials, PID, Council of Scientific and Industrial Research, New Delhi, India, 82-83.
5. Wu LJ, Wang SX, Hua HM (1991). 6-methoxygeniposidic acid, an iridoid glycoside from *Rubia cordifolia*. Phytochemistry 30: 1710.
6. Gorle AM, Patil SS (2010). Evaluation of antioxidant and antiacne property of *Rubia cordifolia*. Der Pharmacia Sinica 1(3): 59-63.
7. Antarkar SS, Chinwalla T, Bhatt N (1983). Anti-inflammatory activity of *Rubia cordifolia* Linn. in rats, Ind J Pharmacol 15: 185-188.
8. Gilani AH, Janbaz KH (1995). Effect of *Rubia cordifolia* extract on acetaminophen and CCl4-induced hepatotoxicity. Phytother Res 9: 372-375.

42

Sida cordifolia Linn.

Family

Malvaceae

Vernacular Name

Sanskrit: Bala; Hindi: Kungyi, Khareti; Bengali: Barila, Brela; English: Country mallow; Gujarati: Jungli methi; Punjabi: Simak; Marathi: Chikana; Malyalam: Kathuram.

Distribution [1]

It is found throughout the India from Himanchal Pradesh to Tamil Nadu, Karnataka, Kerala, West Bengal, Maharashtra, Gujarat, Uttar Pradesh, Assam and Jammu and Kashmir, up to an elevation of 1800 meter.

Traditional Uses [2–3]

It has been used as folk medicine since ancient time. In Ayurveda the whole plant is used as astringent, emollient, tonic and in respiratory diseases. It is also used in rheumatism, colic, haemothermia, leucorrhoea and heart irregularity.

Major Chemical Constituents [4]

β-sitosterol, ephedrine, pseudoephedrine, vasicinone and vasicine.

A Flowering Stalk

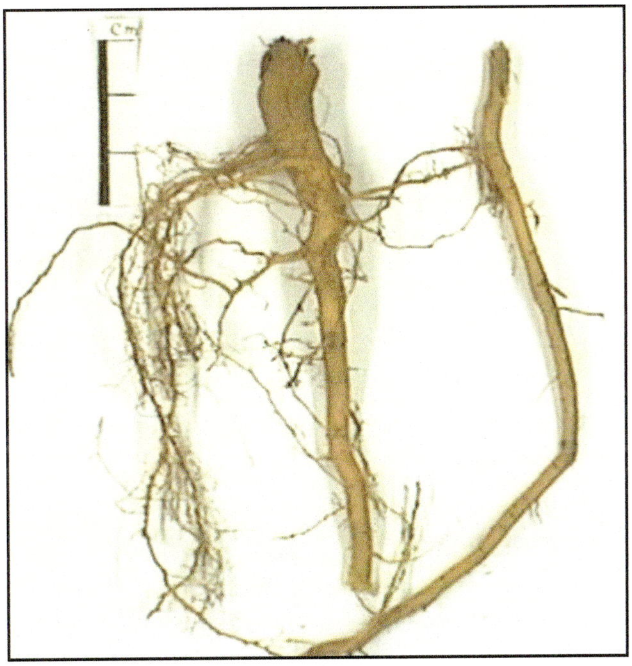

Dried Root

Pharmacological Activity [5-8]

Anti-oxidant, nephroprotective, antibacterial and in Parkinson's disease

Plant Description

It is a perennial shrub, short, erect and woody at base. Leaves are simple, alternate, ovate, ovate-oblong, dentate margin, petiole present, hairy, veins are more on ventral surface. Stem are woody, cylindrical, light brown colored, fibrous, hairy with many branches. Flowers are bisexual, light yellow to cream white colored, axillary without an epicalyx. Roots are adventitious type; cylindrical, woody, grayish colored, smooth with many rootlets are arises from the main root.

Anatomy

Transverse section of root is circular in outline, show narrow range of cork and wide central wood, encircled by narrow phloem and cortex. Detailed structure accounted that the root shows narrow cork made of 4 to 6 row of tangentially elongated dark brown cells; followed by wide parenchymatous cortex filled with few clustered crystals of calcium oxalate as well as starch grains. Phloem narrow, parenchymatous, lignified medullary rays in continuation with of phloem and radially arranged xylem vessels, thin walled fibers and parenchyma. Xylem is well developed, xylem vessels are surrounded by tracheids and xylem fibers.

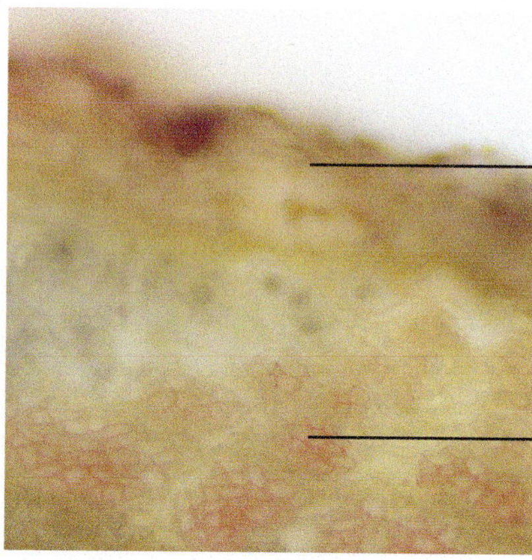

T.S. (40 x)

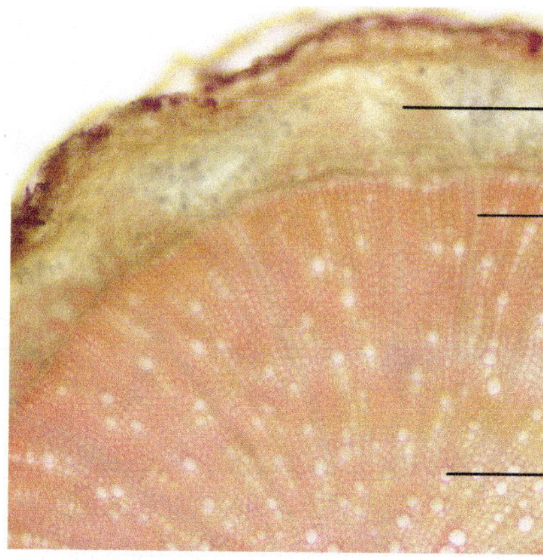

T.S. (10 x)

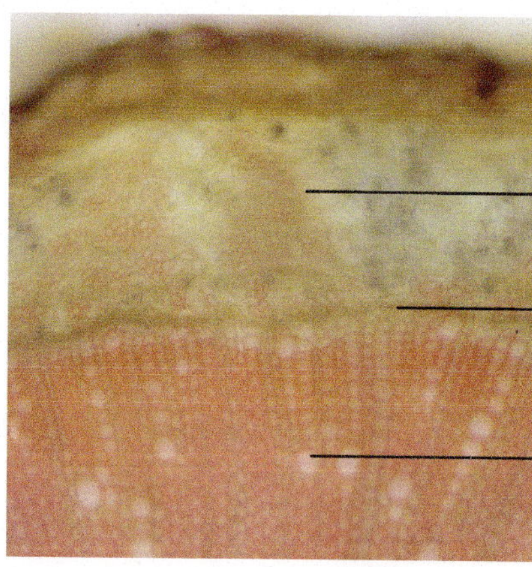

T.S. (20 x)

Transverse Section (T.S.) of *Sida cordifolia* Root.

Abbreviation: ck- cork; cc- calcium oxalate crystal; ct- cortex; mr- medullary rays; ph- phloem; sc- sclereids; xy- xylem.

HPTLC Studies

Working Solution

Accurately weighed (5 g) crude powdered sample was subjected to cold extraction with methanol (50 ml), continuously shaken for 6 h and then allowed to stand at room temperature ($25^{0}C \pm 2$) for 18 h. The extraction was repeated thrice and the pooled filtrate was concentrate in rotatory evaporator (Buchi, Switzerland). Working solution of 10 mg/ml was freshly prepared on the day of quantative analysis.

Standard Solution

β-sitosterol (≥ 98 per cent) was freshly prepared of strength 1 mg/ml.

Sample Application

2 μl of standard solution and 10 μl of sample was applied as 6 mm band with CAMAG HPTLC applicator on activated silica gel 60 GF_{254} HPTLC plate. The chromatogram was developed on pre-saturated CAMAG HPTLC twin trough chamber for 15 min and the solvent system was allowed to run for 8 cm.

Solvent System

Toluene: ethyl acetate: methanol: acetone (7:2:0.2:0.2 v/v)

Identification of β-sitosterol

The chromatogram was visualized after derivatization with aninsaldehyde sulfuric acid reagent and densitometric scanning of β-sitosterol was done at 530 nm.

Identification of β-sitosterol

The standard β-sitosterol was identified at R_f 0.55.

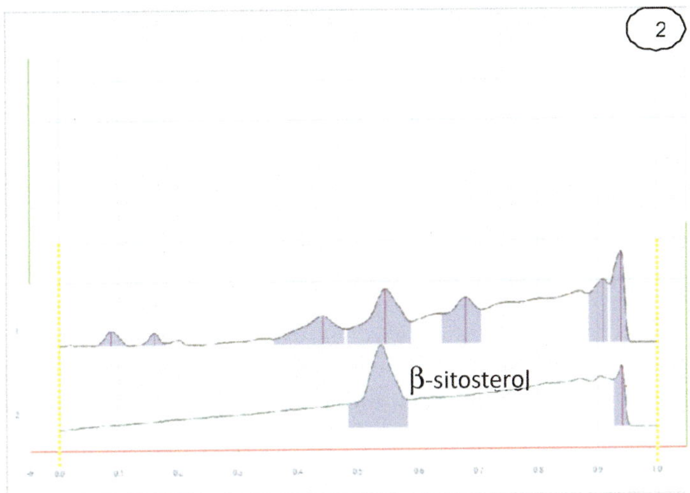

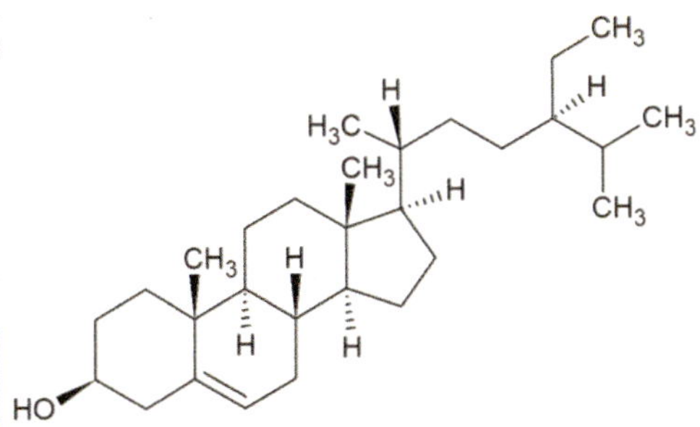

β-sitosterol

Plate 1: HPTLC Chromatogram of *Sida cordifolia* Sample (S) and Standard β-sitosterol (R) in White Light.
Plate 2: Overlay Spectra of Sample and Standard.

1. Chandra K, Chaudhari BG, Dhar BP, Joseph JV, Mangal AK, Dabur R (2007). Database on medicinal plants used in Ayurveda. Cent Counc Res Ayurveda Siddha 8: 42-58.
2. Agharkar SP (1991). medicinal plants of Bombay presidence. Pbl. scientific publishers. Jodhpur, India: 194-195.
3. Singh PA (2006). Bala (*Sida cordifolia*) is it a safe herbal drug? Ethanobotical leaflets 10: 336-342.
4. Ghosal S, Chauhan RRPS, Mehta R (1975). Alkaloids of *Sida cordifolia* L. Phytotherapy Chemistry 14: 830-832.
5. Dhalwal K, Deshpande YS, Purohit AP, Kadam SS (2005). Evaluation of the Antioxidant Activity of *Sida cordifolia*. Pharmaceutical biology 43(9): 754-61.
6. Mehul VM, Nilesh MP, Dharmesh ND, Sarav AD, Bhaskar VH (2012). Assessment of nephroprotective potential of *Sida cordifolia* L. in experimental animals. Scholars Research Library 4 (1): 175-180.
7. Mahesh, Satish (2008). Antimicrobial Activity of Some Important Medicinal Plant against Plant and Human Pathogens. World J Agri Sci 4(S): 839-843.
8. Nagashayana N, Sankarankutty P, Nampoothiri MR, Mohan PK, Mohanakumar KP (2000). Association of L-DOPA with recovery following Ayurveda medication in Parkinson's disease. J Neurol Sci 176(2): 124-7.

43

Sida rhombifolia Linn.

Family

Malvaceae

Vernacular Name

Ayurveda: Mahabala; Sanskrit: Atibala; English: Arrowleaf sida.

Distribution [1]

Sida rhombifolia is mainly found in tropical and warm regions of the world. In India it is found in Tamil Nadu, Karnataka, West Bengal, Maharashtra, Gujarat, Uttar Pradesh and Assam.

Traditional Uses [2–3]

It has been used as folk medicine since ancient time. It is used for the treatment of gonorrhea, gout and rheumatism diuretic, hypertension, diabetes and aphrodisiac.

Major Chemical Constituents [4]

Lupeol, ephedrine, vasicinol, vasicinone and vasicine

Pharmacological Activity [5–7]

Antiproliferative, vasorelaxant, antibacterial and antifungal activity

Plant in Natural Habitat

Dried Piece of Root

Plant Description

It is an annual or perennial herb or shrub, short, erects green, woody at base, simple or glandular hairs over stem. Leaves are simple, alternate, narrowly ovate to lanceolate, entire margin; petiolate, hairy, veins are prominent on both surface. Stem are woody, cylindrical, light brown colored, fibrous and hairy. Flowers are bisexual, solitary or paired, axillary cup-shaped calyx and yellow or white corolla. Roots are adventitious type; cylindrical, woody, white or grayish colored, smooth with many rootlets are arises from the main root.

Anatomy

Transverse section of root is circular in outline, show narrow range of cork and wide central wood, encircled by narrow phloem and cortex. Detailed structure accounted that the root shows narrow cork made of 8 to 10 row of tangentially elongated dark brown cells; followed by wide parenchymatous cortex filled with few clustered crystals of calcium oxalate or sclereids as well as starch grains. Phloem narrow, parenchymatous contains sieve tube, companion cell, fiber and parenchyma. Lignified multiseriate medullary rays in continuation with of phloem and radially arranged with xylem vessels, thin walled fibers and parenchyma. Xylem is well developed and vessels are surrounded by tracheids and xylem fibers.

HPTLC Studies

Working Solution

Accurately weighed (5g) crude powdered sample was subjected to cold extraction with methanol (50 ml), continuously shaken for 6h and then allowed to stand at room temperature (25 ^{0}C ± 2) for 18h. Extraction was repeated thrice, and the pooled filtrate was concentrate in rotatory evaporator (Buchi, Switzerland). Working solution for HPTLC study was made of 10 mg/ml.

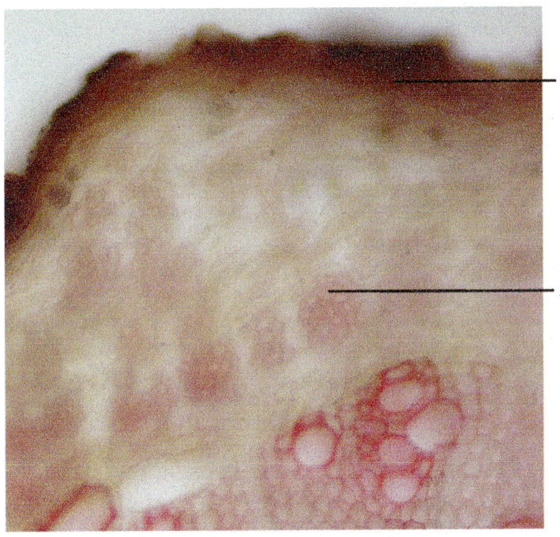

T.S. (10 x)

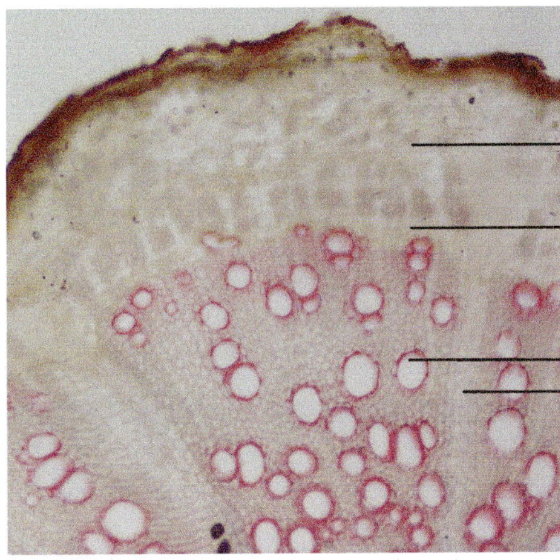

T.S. (4 x)

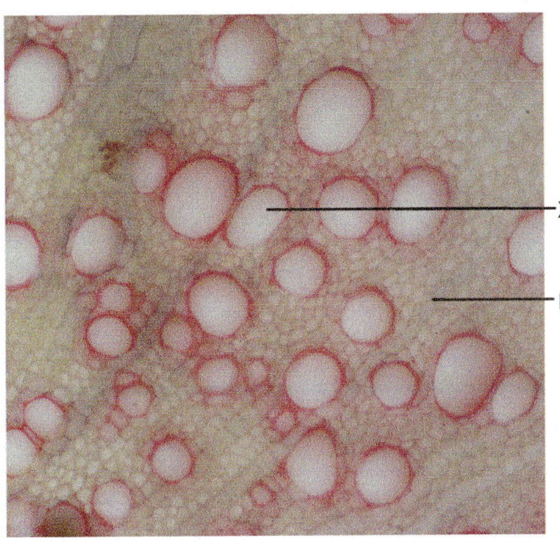

T.S. (20 x)

Transverse Section (T.S.) of *Sida rhombifolia* Root.

Abbreviation: ck- cork; ct- cortex; mr- medullary rays; ph- phloem; sc- sclereids; xy- xylem.

Standard Solution

Lupeol (>98 per cent) of 1 mg/ml concentration was freshly prepared for HPTLC analysis.

Sample Application

2 µl of standard solution and 10 µl of sample was implemented by HPTLC applicator on HPTLC preset silica gel 60 GF254 plate. The 15 minute saturation was given, and the plate was developed in saturated chamber for up to 8 centimetres.

Solvent System

Toluene: ethyl acetate: methanol: acetone (7:2:0.2:0.2 v/v).

Visualization

The chromatogram was visualized after derivatization with anisaldehyde sulphuric acid reagent and densitometric scanning of lupeol was done at 530 nm.

Identification of Lupeol

The standard lupeol was identified at Rf 0.68.

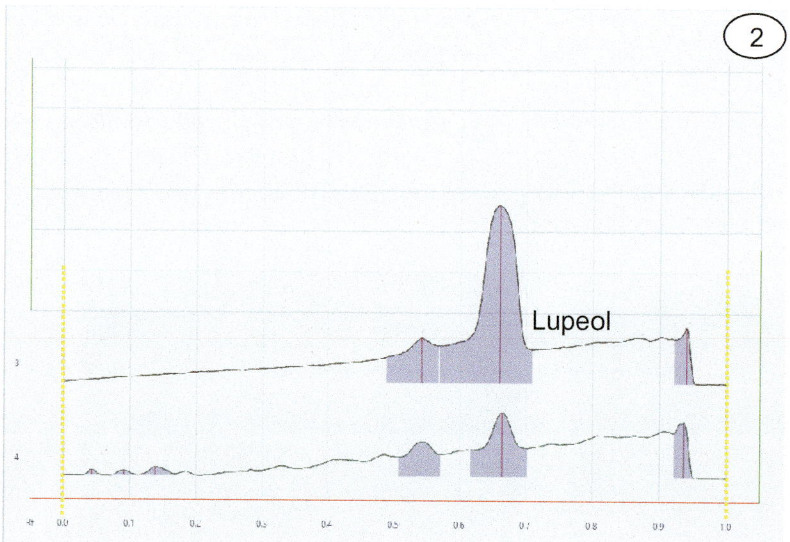

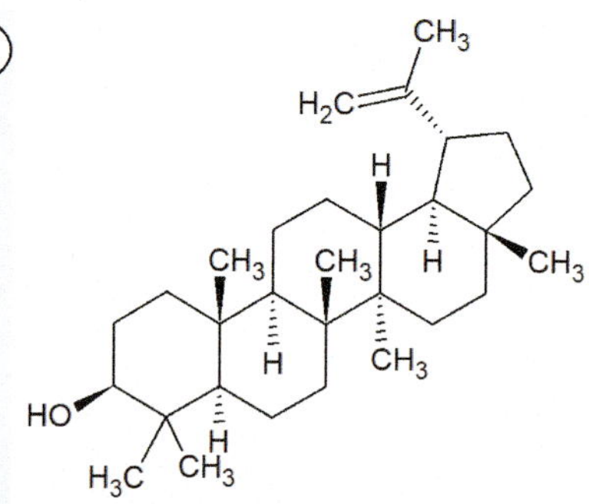

Lupeol

Plate 1: HPTLC Chromatogram of *Sida rhombifolia* Sample (S) and Standard Lupeol (R) in White Light. Plate 2: Overlay Spectra of Sample and Standard.

1. Mabberley DJ (2000). The Plant Book (2nd edn) Cambridge University Press, Cambridge.
2. Nadkarni KM (1982). Indian Materia Medica, 3rd ed., Vol. 3, Bombay Popular Prakashan, Bombay, India 1138.
3. Thounaojam MC, Jadeja RN, Ansarullah, Patel VB, Devkar RV, Ramachandran AV (2009). Potential of *Sida rhomboidea*.Roxb Leaf Extract in Controlling Hypertriglyceridemia in Experimental Models. Pharmacognosy Res 1: 208-212.
4. Prakash A, Verma RK, Ghosal S (1981). Alkaloid constituents of *Sida acuta, S. humilis, S. rhombifolia* and *S. spinosa*. Plant Med 43: 384-388.
5. de Moura CARPES G, Frescura VD, Tedesco SB, Carlos A, DA SILVA F, do CANTO-DOROW TS (2014). Antiproliferative effect of extracts of *Sida rhombifolia* L.(Malvaceae) on the *Allium cepa* cell cycle. Bio Cell 38(1-3): 7-9.
6. Chaves O, Gomes R, Tomaz A, Fernandes M, das Graças Mendes Junior L, de Fátima Agra M, Braga V, de Fátima Vanderlei de Souza M (2013). Secondary metabolites from *Sida rhombifolia* L.(Malvaceae) and the vasorelaxant activity of cryptolepinone. Molecules 18(3): 2769-77.
7. Mishra SH, Chaturvedi SC (1978). Antibacterial and antifungal activity of alkaloid of *Sida rhombifolia*, Indian Drugs 16: 61-63.

44

Solanum incanum Linn.

Family

Solanaceae

Vernacular name

Sanskrit: Barhatam, Sveta-brahti; Hindi: Dholi ringni; English: Bitter apple; Tamil: Karimulli; Telugu: Nidupu vanga.

Distribution [1-2]

This plant is widely available in India, Ethiopia, Uganda, South Africa and Zimbabwe. The species is distributed in dry conditions as a weed along the roadsides and bare field.

Traditional Uses [1-2]

It is used to treat skin problem, stomach disorder, wound, ulcer, in snake bite and chest pain.

Major Chemical Constituents [3]

Solasodine, solamargine, dioscin and protodioscin

Pharmacological Activity [4-5]

Anti-oxidant and anti-plasmodial activity

Plant Description

It is prickly, bright green perennial herb, erect, up to 2 foot in height. Stem is green, hairy, somewhat circular

Plant in Natural Habitat

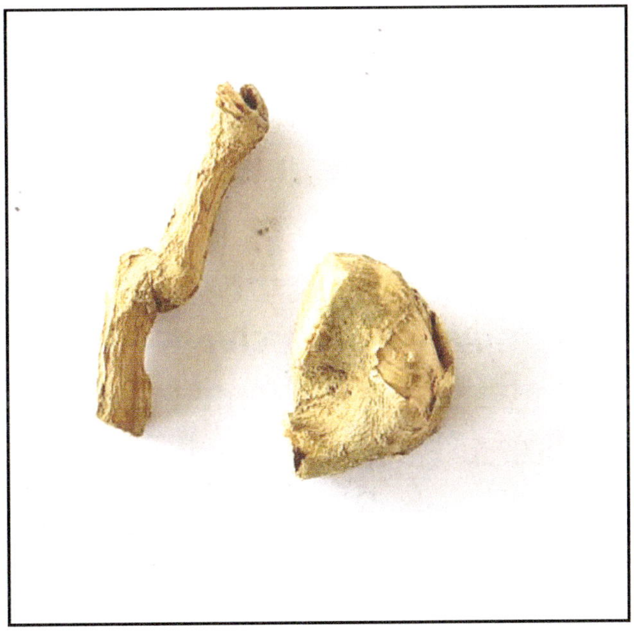

Dried Piece of Roots

and woody at base and branches are numerous and, zigzag; nodes and internodes are clearly seen. Leaves are ovate or elliptic, sub pinnate, hairy on midrib and often on nerves long hairs are present. Flowers are pink or purple coloured, pentamerous with many anthers and ovoid ovary. Berries are green with white strips when unripe and turn yellow when mature, surrounded by enlarged green calyx. Roots are white, smooth, long, cylindrical, thick, woody, covered with a thin friable bark.

Anatomy

Transverse section of mature root is almost circular in outline. The outer layer is made up of rectangular 3 to 5 layered cork cells, followed by 2 or more layered cork cambium. The cortical region is multi layered irregular parenchymatous cells; inner layers are compactly arranged and contain sclereids and starch grains. Sclereids are observed in cortical region, mostly solitary or in a group of 3 to 6. Just below the cortex thick phloem having sieve tubes, companion cells and phloem fibers are present. The xylem is 5-8 celled wide and consists of vessels, fibers, tracheids and xylem parenchyma. Vessels are mostly solitary or in groups of 2-7. The medullary rays are thick-walled, parenchymatous and radially arranged with phloem and xylem.

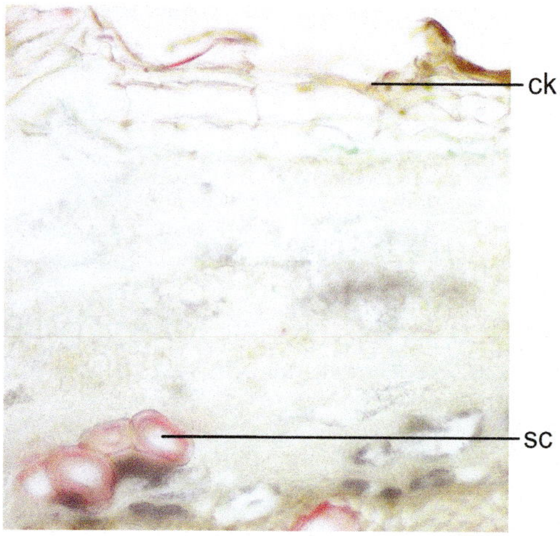

T.S. (20 x)

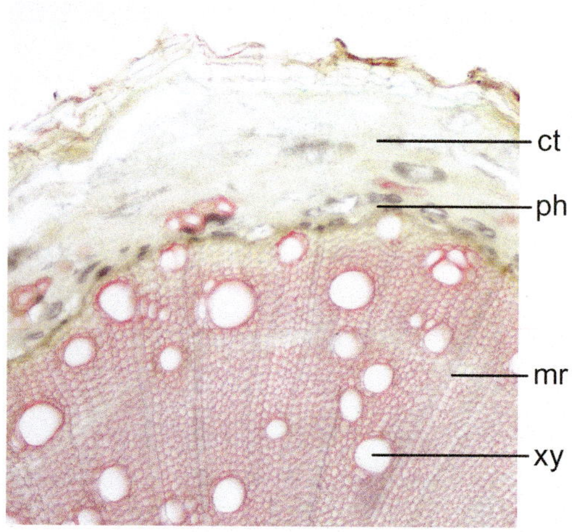

T.S. (10 x)

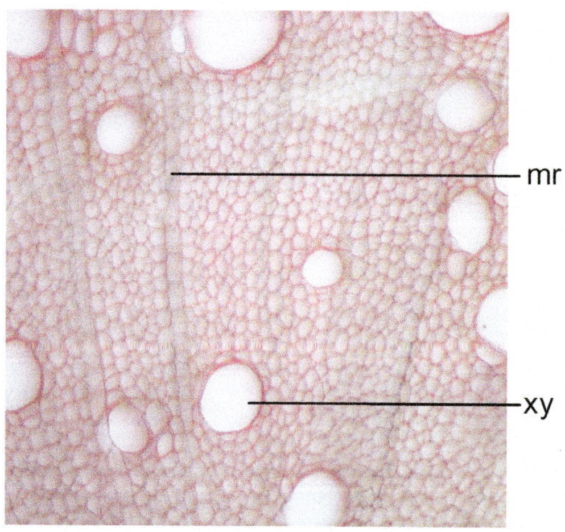

T.S. (20 x)

Transverse Section (T.S.) of *Solanum incanum* Root.

Abbreviation: ck- cork; ct- cortex; mr- medullary rays; ph- phloem; sc- sclereids; xy- xylem.

HPTLC Studies

Sample Solution

Accurately weight (5 g) crude powdered sample was hydrolyzed with (50 ml) 2.5 N hydrochloric acid for 4 hrs at 80 °C. The residue obtained was filtered, washed with hot water until neutral or free from acid and further dried hot air oven. The hydrolyzed residue was further extracted with methanol (50 ml) followed by continuously shaken for 6 h and then allowed to stand for 18 h at room temperature (25 °C ± 2). Extraction was repeated thrice, filtered and the pooled filtrate was concentrate in rotatory evaporator (Buchi, Switzerland), dried residue was stored at 4 °C. Working solution of strength 10 mg/ml was freshly prepared on the day of analysis.

Standard Solution

Solasodine (≥ 98 per cent) was freshly prepared of strength 1 mg/ml.

Sample Application

2 µl of standard solution and 10 µl of sample was applied as 6 mm band with CAMAG HPTLC applicator on activated silica gel 60 GF_{254} HPTLC plate. The chromatogram was developed on pre-saturated CAMAG HPTLC twin trough chamber for 20 min and the solvent system was allowed to run for 8 cm.

Solvent System

Toluene- ethyl acetate- diethyl amine (6:2:0.3 v/v)

Visualization

The chromatogram was visualized after derivatization with anisaldehyde sulphuric acid reagent and densitometric scanning of solasodine was done at 540 nm.

Identification of Solasodine

The standard solasodine was identified at R_f 0.32.

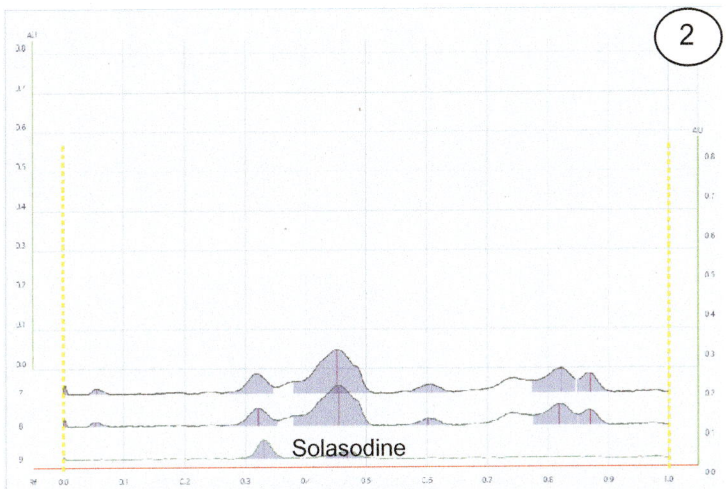

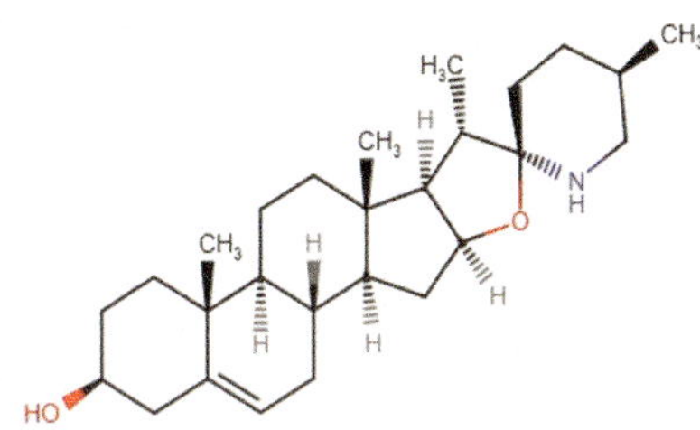

Solasodine

Plate 1: HPTLC Chromatogram of *Solanum incanum* Sample (S) and Standard Solasodine (R) in White Light.

Plate 2: Overlay Spectra of Sample and Standard.

1. Giday M, Asfaw Z, Elmqvist T, Woldu Z (2003). "An ethnobotanical Study of Medicinal Plants Used by Zay People in Ethiopia", J Ethnophar 85: 43-52.
2. Hyde M, Wursten B (2006). "Flora of Zimbabwe", Species Information of *Solanum incanum*. http://www.zimbabweflora.co.zw/speciesdata/species-display.php?species_id=150670.
3. Manase, M.J., Mitaine-Offer, A.C., Pertuit, *et al., Solanum incanum* and *S. heteracanthum* as sources of biologically active steroid glycosides: confirmation of their synonymy. Fitoterapia., 83(6):1115-9 (2012).
4. Makler MT, Ries JM, Williams JA, Bancroft JE, Piper RC, Hinrichs DJ (1993). Parasite lactate dehydrogenase as an assay for Plasmodium falciparum drug sensitivity. Am J Trop Med Hyg 48: 739-741.
5. Anwar S (2018). Pharmacological Investigation of *Solanum incanum* against P. falciparum, L. infantum, T. cruzi and T. brucei: A Role of Antioxidant Effect and Clinical Overview. Bio Pharma J 11(2): 653-60.

45

Solanum indicum Linn.; Syn: *Solanum anguivi* Lam.

Family

Solanaceae

Vernacular name [1]

Sanskrit: Brihati; Hindi: Badi kateli, Badi bhatkataiya; Urdu: Kateli; Punjabi: Kandiarivaddi; Malayalam: Cheru.

Distribution [1]

The plant is widely spread on roadside as well as in wasteland of warmer and dry part of the country up to an altitude of 1500 m.

Traditional Uses [1–3]

The whole plant is used in various diseases such as asthma, dysuria, bronchitis, leucoderma and cardiac weakness. In Rajasthan root is used for aphrodisiac potential.

Major Chemical Constituents [4–5]

Solasodine, solamargine and scopoltin

Pharmacological Activity [6–8]

Anti-pyretic, anti-inflammatory, anti-bacterial and anti-cancer activity.

Plant in Natural Habitat

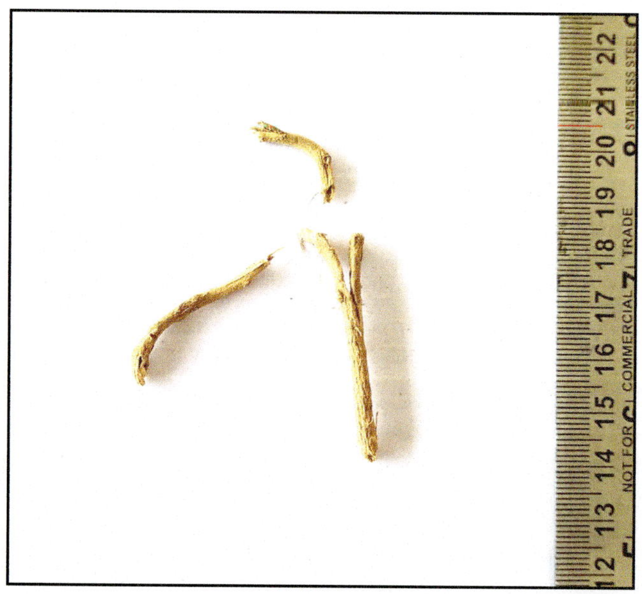

Dried Roots

Plant Description

It is green perennial shrub, erect, up to 2 to 5 foot in height. Stem is somewhat circular, woody at base, hairy, green, numerous zigzag branches, stout, and prickly, nodes and internodes are clearly seen. Leaves are ovate or elliptic, sub pinnate, pointed apex and stellately woolly beneath, on midrib and often on nerves long hairs are present. Flowers are blue coloured, axillary raceme pentamerous with many anthers and ovoid ovary. Berries are green, rounded and surrounded by enlarged green calyx. Roots are white, long, cylindrical, thick, woody, numerous secondary roots arise, fracture rough due to root scars.

Anatomy

Transverse section of mature root is almost circular in outline. The outer layer is made up of rectangular 8 to 10 layered cork cells, followed by 2 or more layered cork cambium. The cortical region is multi layered irregular parenchymatous cells; calcium oxalate crystals and starch grains are frequently present. Calcium crystals are observed in cortical region, mostly solitary or in a group of 3 to 10 or more. Just below the cortex thick phloem having sieve tubes, companion cells and phloem fibers are present. The xylem is wide and consists of vessels, fibers, tracheids and xylem parenchyma. Vessels are mostly solitary or in groups of 2-8. The lignified medullary rays are radially arranged with parenchymatous phloem and xylem.

HPTLC Studies

Sample Solution

Accurately weight (5 g) crude powdered sample was hydrolyzed with (50 ml) 2.5 N hydrochloric acid for 4 hrs at 80 ^{o}C. The residue obtained was filtered, washed with hot water until neutral or free from acid

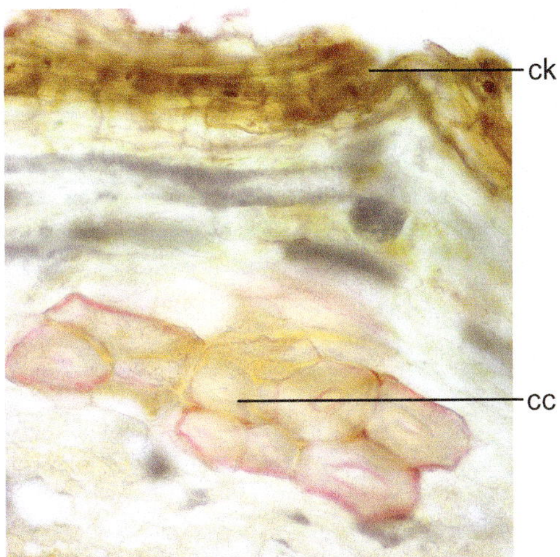

T.S. (20 x)

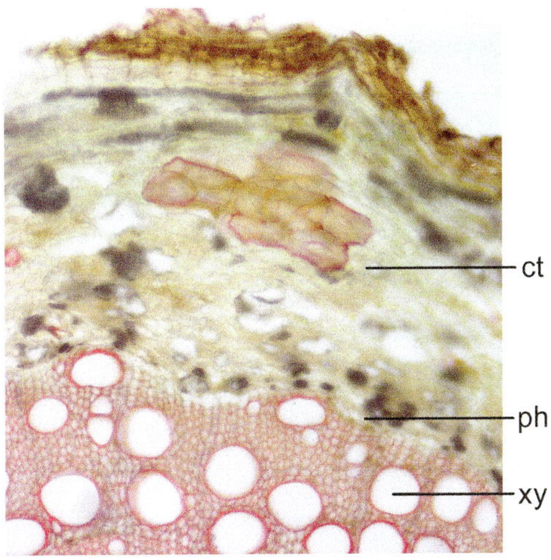

T.S. (10 x)

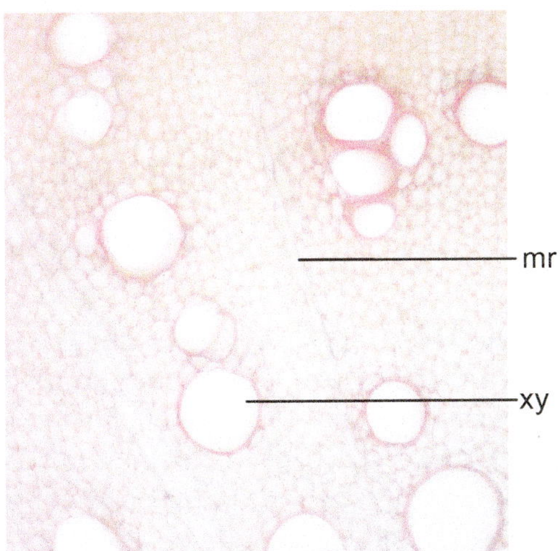

T.S. (20 x)

Transverse Section (T.S.) of *Solanum indicum* Root.

Abbreviation: ck- cork; ct- cortex; mr- medullary rays; ph- phloem; cc- calcium oxalate crystal; xy- xylem.

and further dried hot air oven. The hydrolyzed residue was further extracted with methanol (50 ml) followed by continuously shaken for 6 h and then allowed to stand for 18 h at room temperature (25 °C ± 2). Extraction was repeated thrice, filtered and the pooled filtrate was concentrate in rotatory evaporator (Buchi, Switzerland), dried residue was stored at 4 °C. Working solution of strength 10 mg/ml was freshly prepared on the day of analysis.

Standard Solution

Solasodine (≥ 98 per cent) was freshly prepared of strength 1 mg/ml.

Sample Application

2 µl of standard solution and 10 µl of sample was applied as 6 mm band with CAMAG HPTLC applicator on activated silica gel 60 GF_{254} HPTLC plate. The chromatogram was developed on pre-saturated CAMAG HPTLC twin trough chamber for 20 min and the solvent system was allowed to run for 8 cm.

Solvent System

Toluene: ethyl acetate: diethyl amine (6:2:0.3 v/v)

Visualization

The chromatogram was visualized after derivatization with anisaldehyde sulphuric acid reagent and densitometric scanning of solasodine was done at 540 nm.

Identification of Solasodine

The standard solasodine was identified at R_f 0.32.

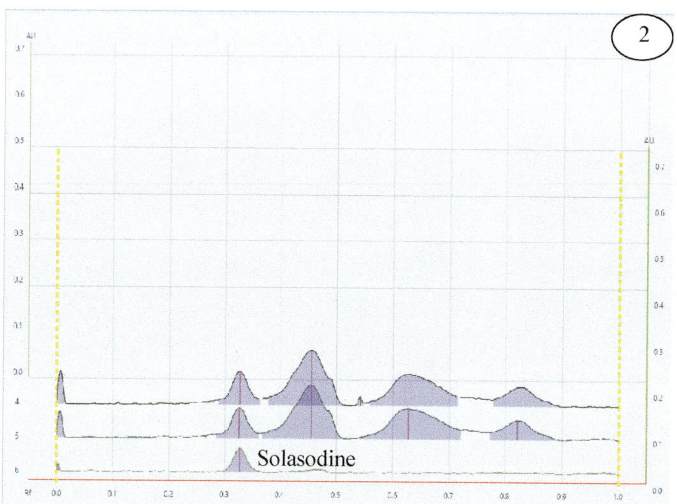

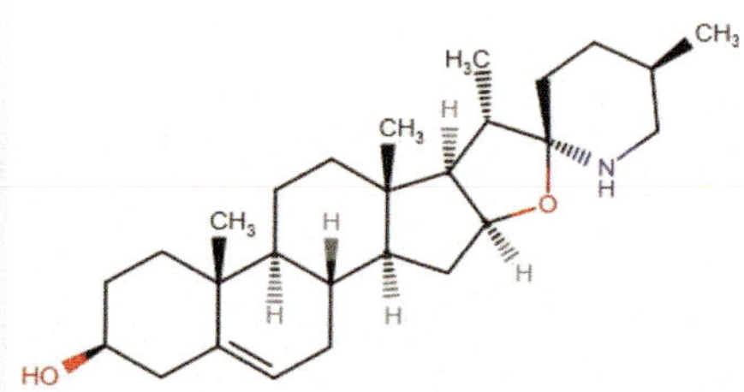

Solasodine

Plate 1: HPTLC Chroamatogram of *Solanum indicum* (s) and Standard Solasodine (R) in White Light.

Plate 2: Overlay Spectra of Sample and Standard.

1. Anonymous (1999).The Ayurvedic Pharmacopoeia of India Part-I Volume – II Government of India, Ministry of Health and Family Welfare Department of AYUSH 28.

2. Bhakta T (2004). Common Vegetables of the Tribals of Tripura. Agartala, Tripura, India. Tripura Tribal Research Institute.

3. Gupta RB, Ahuja A, Sharma N, Kabra MP (2013). Indigenous Herbal Plants used by tribes of Rajasthan; Improving Sexual Performance and Problem of Sexuality. Int J Drug Develop Res 5(2): 40-46.

4. Syu W, Don M, Lee G, Sun C (2001). Cytotoxic and novel compounds from *Solanum indicum*. J Nat Prod 64(9): 1232-3.

5. Yahara S, Nakamura T, Someya Y, Matsumoto T, Yamashita T, Nohara T (1996). Steroidal glycosides indiosides A–E, from *Solanum indicum*. Phytochem 43(6): 1319-23.

6. Gavimath CC, Kulkarni SM, Raorane CJ, Gavade BG, Kalsekar DP, Ravishankar BE, Hooli RS (2012). Antibacterial potentials of *Solanum indicum* Linn, *Solanum xanthocarpum* and *Physalis minima*. Int J Pharma App 3(4): 414-8.

7. Ma P, Cao TT, Gu GF, Zhao X, Du YG, Zhang Y (2006). Inducement effect of synthetic indiosides from *Solanum indicum* L. on apoptosis of human hepatocarcinoma cell line Bel-7402 and its mechanism. Ai Zheng 25(4): 438-42.

8. Deb P K, Ghosh R, Chakraverty R, Debnath R, Das L, Bhakta T (2014). Phytochemical and pharmacological evaluation of fruits of *Solanum indicum* Linn. Int J Pharm Sci Rev Res 25(2): 28-32.

46

Solanum virginianum L. Syn. *Solanum xanthocarpum* Schrad. and Wendl.

Family

Solanaceae

Vernacular name [1]

Sanskrit: Kantkari; Hindi: Kateri, Bhatakataiya; Bengali: Kantakari; English: Febrifuge Plant, Yellow berried night shade; Gujarati: Bhoringni; Punjabi: Kandiari.

Distribution [2]

It is found throughout India in dry and humid places as a weed along the roadsides and bare fields.

Plant in Natural Habitat

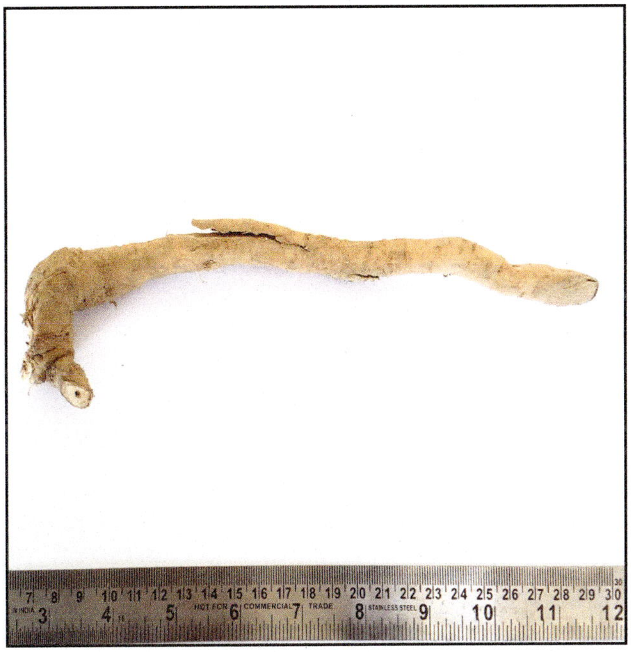

Fresh Root

Traditional Uses [3-4]

In Ayurveda, Kantkari is described as bitter, pungent, digestive and carminative. Decoction of roots is used as diuretic, expectorant and is one of the key constituent of "Dasamoola". The plant extract (entire) is used in bronchial asthma, tympanitis, piles and dysuria. Kantkari Ghrita of Charaka is specific for cough and asthma.

Major Chemical Constituents [5-6]

Solasodine, solamargine, diosgenin and β-Sitosterol.

Pharmacological Activity [7-10]

Anti-fertility, anti-diabetic, hepatoprotective and anti-asthmatic activity.

Plant Description

It is very prickly bright green perennial herb, prostrate or erect up to 1 foot. Stem is somewhat circular and woody at base, small prickles are present all over the aerial part, branches are numerous and zigzag; nodes and internodes are clearly seen. Leaves are ovate or elliptic, sub pinnate, hairy, on midrib and often on nerves long shiny prickles are present. Flowers are purple coloured, pentamerous with many anthers and ovoid ovary. Berries are green with white strips when unripe and yellow when mature and surrounded by enlarged green calyx. Roots are white, smooth, long, cylindrical, thick, woody, covered with a thin friable bark.

Anatomy

Transverse section of root is circular in outline, show narrow range of cork and central wood, encircled by narrow phloem and wider cortex. Detailed structure accounted that the root shows narrow cork made of 2 to 4 row of dark brown cells, followed by wide parenchymatous cortex, filled with small surrounded starch grains. Phloem narrow, parenchymatous, made of group of fibers and cambium distinct. Lignified medullary rays in continuation with phloem and radially arranged xylem, thin walled fibers. Xylem composed of xylem vessels, tracheids and fibers.

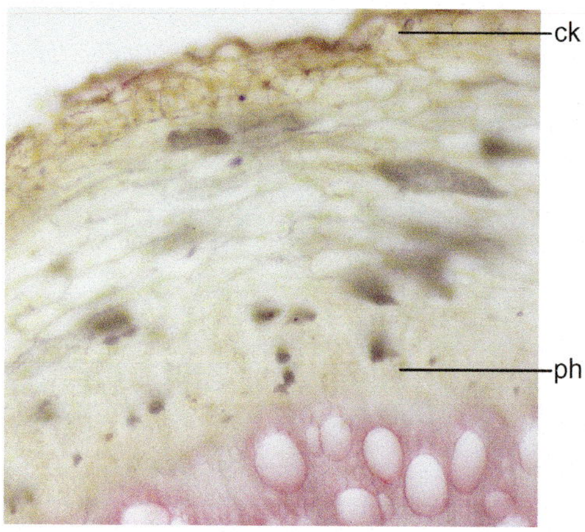

T.S. (10 x)

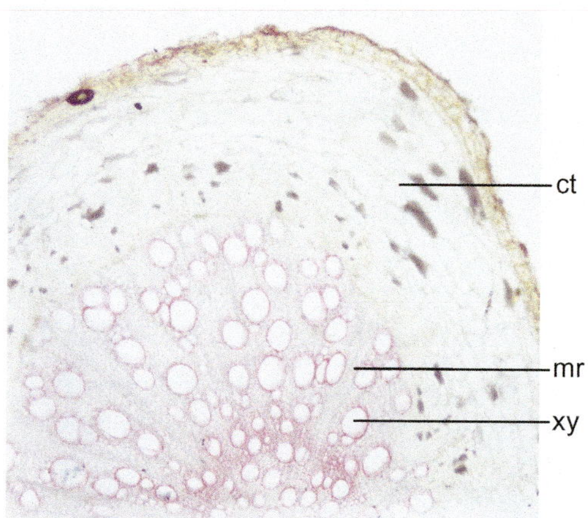

T.S. (4 x)

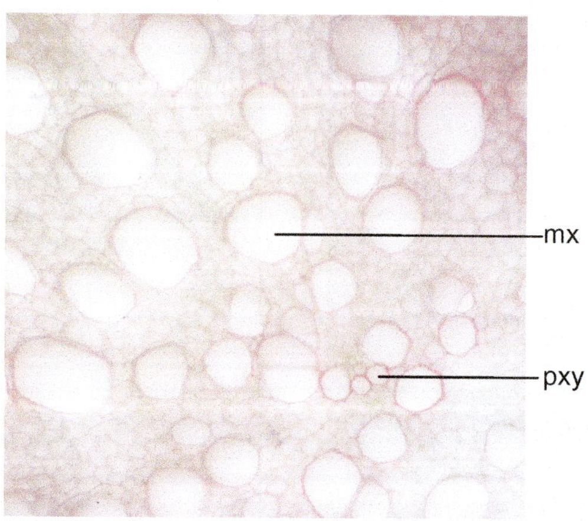

T.S. (20 x)

Transverse Section (T.S.) of *Solanum xanthocarpum* Root.

Abbreviation: ck- cork; ct- cortex; mr- medullary rays; mx- meta xylem; ph- phloem; pxy- proto xylem; xy- xylem.

HPTLC Studies

Sample Solution

Accurately weight (5 g) crude powdered sample was hydrolyzed with (50 ml) 2.5 N hydrochloric acid for 4 hrs at 80 °C. The residue obtained was filtered, washed with hot water until neutral or free from acid and further dried hot air oven. The hydrolyzed residue was further extracted with methanol (50 ml) followed by continuously shaken for 6 h and then allowed to stand for 18 h at room temperature (25 °C ± 2). Extraction was repeated thrice, filtered and the pooled filtrate was concentrate in rotatory evaporator (Buchi, Switzerland), dried residue was stored at 4 °C. Working solution of strength 10 mg/ml was freshly prepared on the day of analysis.

Standard Solution

Solasodine (≥ 98 per cent) was freshly prepared of strength 1 mg/ml.

Sample Application

2 µl of standard solution and 10 µl of sample was applied as 6 mm band with CAMAG HPTLC applicator on activated silica gel 60 GF_{254} HPTLC plate. The chromatogram was developed on pre-saturated CAMAG HPTLC twin trough chamber for 20 min and the solvent system was allowed to run for 8 cm.

Solvent System

Toluene- ethyl acetate- diethyl amine (6:2:0.3 v/v)

Visualization

The chromatogram was visualized after derivatization with anisaldehyde sulphuric acid reagent and densitometric scanning of solasodine was done at 540 nm.

Identification of Solasodine

The standard solasodine was identified at R_f 0.32.

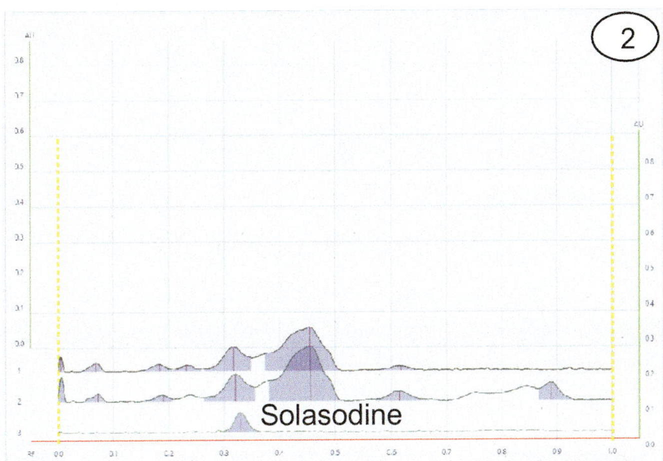

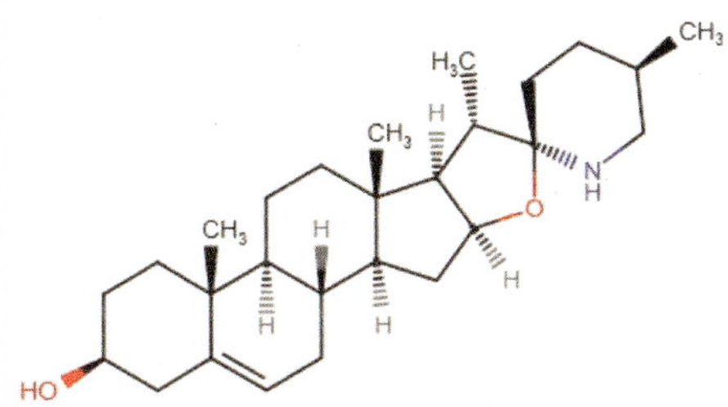

Solasodine

Plate 1: HPTLC Chromatogram of *Solanum xanthocarpum* (S) and Standard Solasodine (R) in White Light.

Plate 2: Overlay Spectra of Sample and Standard.

1. Anonymous (2001). The Ayurvedic Pharmacopoeia of India, Part-I, Vol. I, 1st ed. New Delhi: Govt. of India, Ministry of Health and Family Welfare, Dept. of I.S.M. and H 59-60.
2. Sheth AK (2005). The Herbs of Ayurveda. A.K. Sheth Publisher,Vol.IV: 1044.
3. Ghani A (1998). Medicinal plants of Bangladesh - chemical constituents and uses. Asiatic Society of Bangladesh, Dhaka.
4. Khare CP (1995). Encyclopedia of Indian Medicinal Plants, Springer: 432-433.
5. Heble MR, Narayanaswami S, Chadha MS (1968). Diosgenin and β-Sitosterol: Isolation from *Solanum xanthocarpum* Tissue Cultures. AAASS 161(13): 1145.
6. Dixit VP (1980). Antifertility effects of Solasodine obtained from *Solanum xanthocarpum* berries in male rats and dogs. Department of Zoology, University of Rajasthan, Jaipur, India.
7. Gupta S, Mal M, Bhattacharya P (2005). Eur Bull Drug Res 13: 51-55.
8. Najmi AK, Pillai KK, Pal SN, Aqil M (2005). Free radical scavenging and hepatoprotective activity of jigrine against D-GalN induced hepatopathy in rats. J Ethnopharmacol 97: 521-525.
9. Vadnere GP, Gaud RS, Singhai AK (2008). Evaluation of Anti-Asthmatic Property of *Solanum xanthocarpum* Flower Extracts. Pharmacology online 1: 513-522.

47

Tephrosia purpurea (L.) Pers.

Family

Fabaceae

Vernacular name

Sanskrit: Sharpunkha; Hindi: Sarphonk, Sharpunkha; Rajasthani: Masa; Gujarati: Unhali; English: Fish poison, Wild indigo.

Distribution [1]

Tephrosia purpurea is widely distributed in Asian countries like China, India and Sri Lanka and also in Australian continent.

Traditional Uses [2–3]

It is used in folk medicine to treat jaundice, bronchitis, pimples, labor pain, cough and cold, pyrrhia and all types of wounds.

Major Chemical Constituents [4–5]

Lupeol, rutin, β-sitosterol, quercetin and tephrosin

Pharmacological Activity [2, 6–8]

Anti-oxidant, anti-ulcer, hepatoprotective and anti-inflammatory activity

A Flowering Plant

Dried Root

Morphological Description

Tephrosia purpurea is an annual or perennial short lived, erect or spreading herb growing up to one meter. Stem is woody, erect, cylindrical and containing a pair of spine at nodes. Leaves are large, imparipinnate, opposite, 8 to 12 pairs leaflet with terminal one. Leaflets are small, opposite, sessile or petiolulate, margin smooth and prominent veins (netted) on both sides. Inflorescence is recemose, flowers are small, purple or red, bisexual, zygomorphic and seeds are ellipsoidal. The root is smooth, long, cylindrical, tapering, brownish yellow in colour, characteristic odour and bitter taste.

Anatomy

Transverse section of root is circular in outline, show very narrow range of cork, wider cortex and well developed wide stellar system. Detailed structure accounted that the root shows narrow cork, followed by multi-layered thin-walled cork cambium lie underneath this. The cortical region is made up of 15 to 18 layered irregular parenchymatous cells, containing round or oval starch grains. Just below the cortex phloem are present having sieve tubes, companion cells and phloem fibers. The xylem is wide and consists of vessels, fibers, tracheids and xylem parenchyma. Vessels are mostly in groups of 2 to 5 and rarely solitary. The medullary rays are thick-walled, multiseriate, radiating, and parenchymatous containing starch grains. In longitudinal section vessels are seen vertically running broad segmented canals with bordered pitted and thickened wall, parenchyma are seen as narrow elongated tubes. The arrangements of medullary rays are horizontal, along with xylem vessels and fibers; tracheids are found along with the xylem vessels.

HPTLC Studies

Sample Solution

Accurately weight (2 g) crude powdered sample was extracted (cold) with methanol (50 ml) followed by continuously shaken for 6 h and then allowed to stand for 18 h at room temperature (25 ± 2°C). The extraction was repeated thrice, filtered and the pooled filtrate was concentrate in rotatory evaporator (Buchi, Switzerland), dried residue was stored at 4 °C. Working solution of strength 10 mg/ml was freshly prepared on the day of quantative analysis.

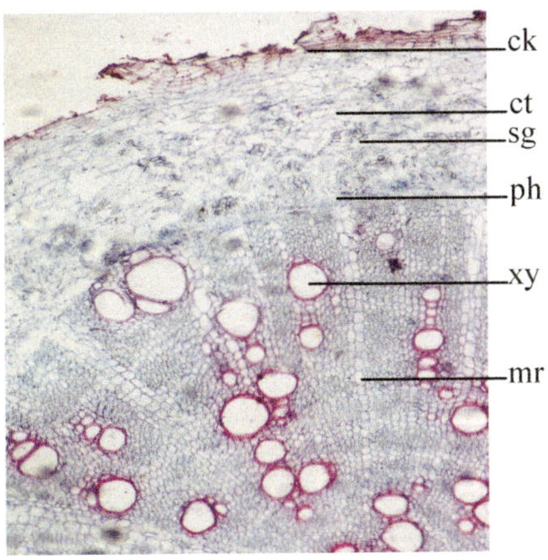

T.S. (10 x)

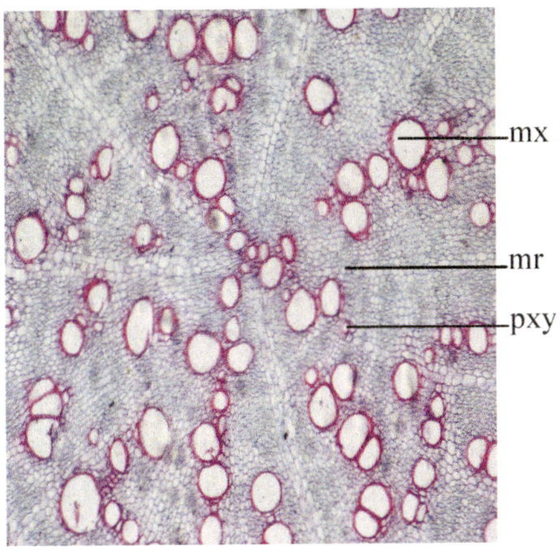

T.S. (10 x)

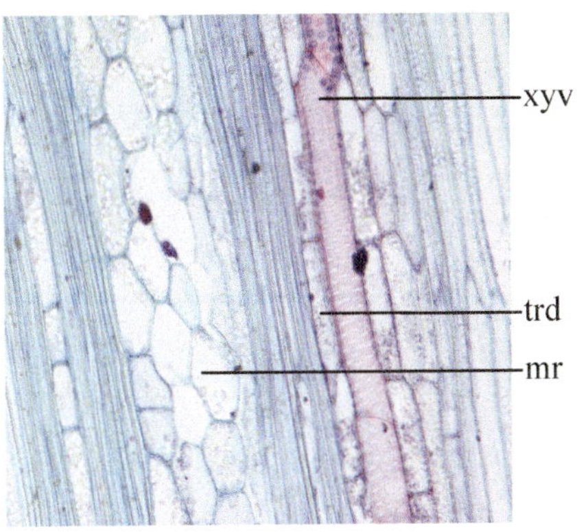

L.S. (40 x)

Transverse Section (T.S.) & Longitudinal Section (L.S.) of *Tephrosia purpurea* Root.

Abbreviation:; ck- cork; ct- cortex; mr- medullary rays; mx- meta xylem; ph- phloem; pxy- proto xylem; sg- starch grains; trd- tracheid; xy- xylem; xyv- xylem vessels.

Standard Solution

Lupeol (≥ 98 per cent) was freshly prepared of strength 1 mg/ml.

Sample Application

2 µl of standard solution and 10 µl of sample was applied as 6 mm band with CAMAG HPTLC applicator on activated silica gel 60 GF_{254} HPTLC plate. The chromatogram was developed on pre-saturated CAMAG HPTLC twin trough chamber for 20 min and the solvent system was allowed to run for 8 cm.

Solvent System

Toluene: ethyl acetate: methanol: formic acid (7:2:0.5:0.5 v/v).

Visualization

The chromatogram was visualized after derivatization with anisaldehyde sulphuric acid reagent and densitometric scanning of lupeol was done at 530 nm.

Identification of Lupeol

The standard lupeol was identified at R_f 0.63.

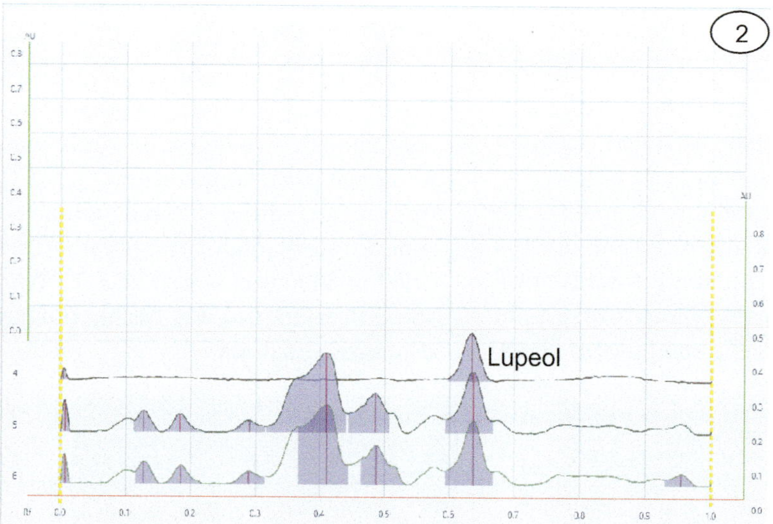

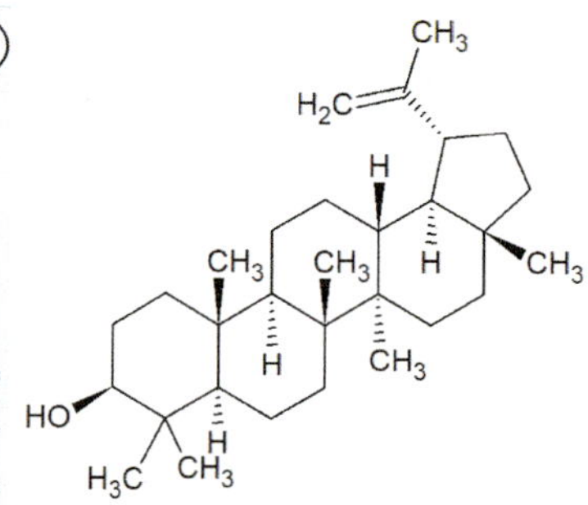

Lupeol

Plate 1: HPTLC Chromatogram of *Tephrosia purpurea* Sample (S) and Standard Lupeol (R) in White Light.
Plate 2: Overlay Spectra of Sample and Standard.

1. Akanksha B, Avijit M, Chakraborthy GS, Seema G (2014). Phytopharmacological uses of *Tephrosia purpurea*-A review. Pharmacophore 5(4):658-65.

2. Kumar CH, Ramesh A, Kumar JNS (2011). A review of hepatoprotective activity of medicinal plants [J]. Int J Pharm Sci Res 2(3): 501-515.

3. Billore KV, Yelne MB, Dennis TJ (2004). Database on Medicinal Plants Used in Ayurveda, [M]. Vol. 6, Central Council for Research in Ayurveda and Siddha, India: New Delhi 358-373.

4. Kumar M, Gehlot S (2012). "Systemic effect of *Tephrosia purpurea* (Sarapunkha) on G.I.T.-An Experimental Study". Int J Ayu Herbal Med 2: 328-335.

5. Mohamed EFH, Mohamed H, Abd ER (2009). Rare prenylated flavonoids from *Tephrosia purpurea* [J]. Phytochemistry 70(12): 1472-1474.

6. Soni K, Kumar PS, Saraf MN (2006). Antioxidant activity of *Tephrosia purpurea* [J]. Indian J Pharm Sci 68(4): 456-460.

7. Deshpande SS, Shah GB (2003). Pharmacological activity of *Tephrosia purpurea*. American Association Pharmaceuti Sci J 10:S2.

8. Gopalakrishnan S, Vadivel E, Dhanalakshmi K (2010). "Antiinflammatoty and analgesic activities of *Tephrosia purpurea* Linn. Aerial and root extracts". J Pha Res 3 (5): 1103-1106.

48

Thalictrum foliolosum DC.

Family

Ranunculaceae

Vernacular Name [1]

Sanskrit: Piyaranga, Trayamana; Hindi: Macharmar, Pilijari; Tamil: Citarokani, Inkukati; Marathi: Garbeeni, Mameeri.

Distribution

Thalictrum foliolosum is distributed in India, Nepal and Bhutan at an altitude of 1000-3400 m. In India, the species is found in Jammu and Kashmir, Himachal Pradesh, Sikkim, Arunachal Pradesh, Meghalaya, Orissa and Tamil Nadu.

Traditional Uses [2-3]

Traditionally the root is used as a tonic, febrifuge, diuretic, antiseptic as well as in treatment of flatulence, jaundice and rheumatism. Decoction of root is used as antidote in snake bite.

Major Chemical Constituents [4]

Berberine, Palmitine and jatrorrhizine.

Plant in Natural Habitat

Dried Root Powder

Major Pharmacological Activity [5-6]

Anti-oxidant, anti-microbial and anti-malarial activity.

Plant Description

Thalictrum foliolosum is a large perennial herb attaining height about 1-1.3 meter. Leaves are large, petiolate, leaflets are rhombic-elliptic or ovate, leathery, margin dentate with prominent veins on both side. Stem is long, erect, cylindrical, soft sometime woody, glberous and branched. Inflorescence is terminal or axillary with many tiny yellowish flowers. Roots are woody, cylindrical, outer brownish brittle bark and internal yellow wood.

Anatomy

Transverse section is circular in outline. The outer layer is made up of rectangular 4 to 6 layered cork cells, followed by 2 or 3 layered cork cambium. The cortical region is made up of 30 to 35 layered irregular parenchymatous cells; outer cortical cells are broad having intercellular spaces, smaller towards central region. Just below the cortex 4 to 10 layered thick phloem having sieve tubes, companion cells and phloem fibers are present. The xylem is distributed throughout the vascular region, and consists of vessels, fibers, tracheids and xylem parenchyma. The xylem vessels are solitary or in group of 2-3 and center are occupied by vessels and tracheids.

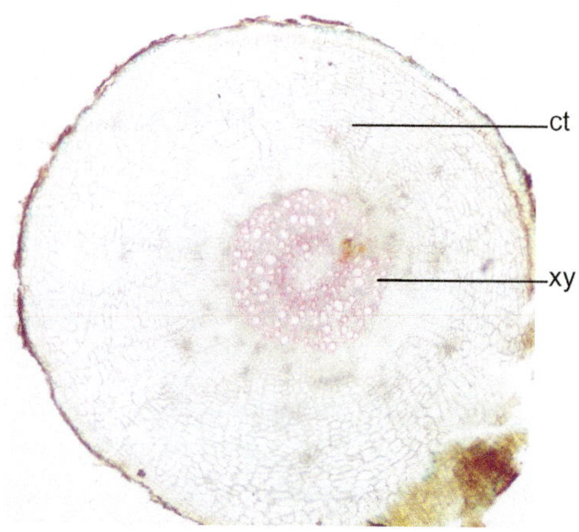

T.S. (4 x)

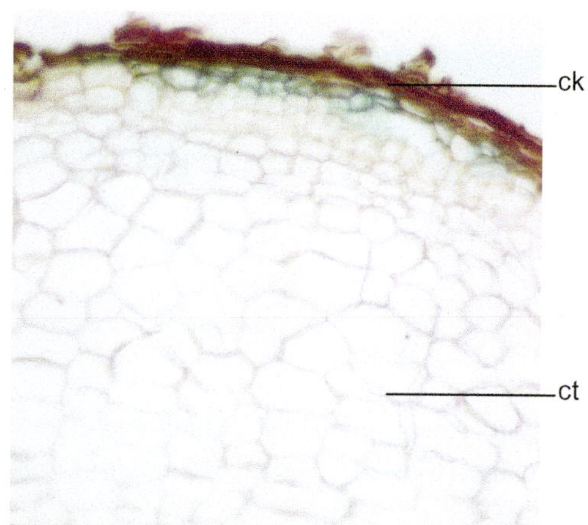

T.S. (20 x)

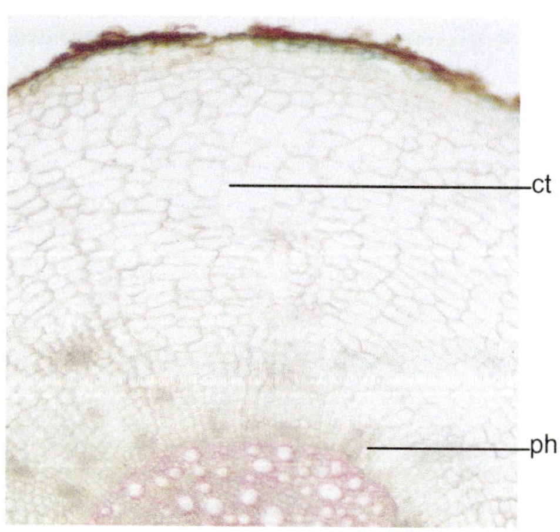

T.S. (10 x)

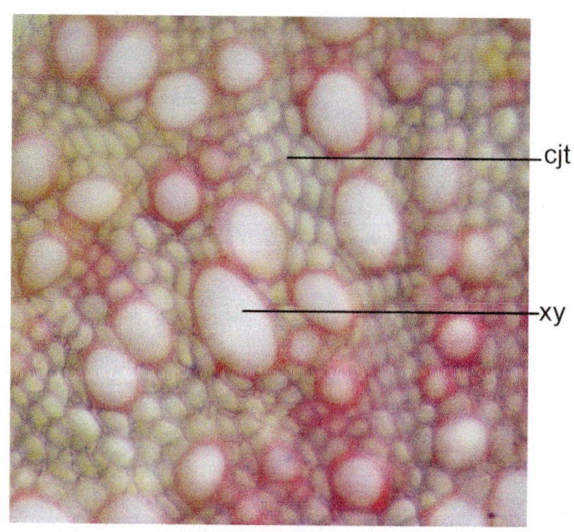

T.S. (20 x)

Transverse Section (T.S.) of *Thalictrum foliolosum* Root.

Abbreviation: ck- cork; ct- cortex; cjt- conjunctive tissue; ph- phloem; xy- xylem.

HPTLC Studies

Sample Solution

Accurately weight (5 g) crude powdered sample was subjected to cold extraction with methanol (50 ml), continuously shaken for 6 h and then allowed to stand at room temperature (25 ^{0}C ± 2) for 18 h. The extraction was repeated thrice, filtered and the pooled filtrate was concentrate in rotatory evaporator (Buchi, Switzerland). Working solution of strength 10 mg/ml was freshly prepared on the day of quantative analysis.

Standard Solution

Berberine (≥ 98 per cent) was freshly prepared of strength 1 mg/ml.

Sample Application

2 μl of standard solution and 10 μl of sample was applied as 6 mm band with CAMAG HPTLC applicator on activated silica gel 60 GF_{254} HPTLC plate. The chromatogram was developed on pre-saturated CAMAG HPTLC twin trough chamber for 30 min and the solvent system was allowed to run for 8 cm.

Solvent System

n-propano- water- formic acid (9:1:0.1 v/v).

Visualization

The chromatogram was visualized at UV_{366} and densitometric scanning of berberine was done at 350 nm.

Identification of Berberine

The standard berberine was identified at R_f 0.32.

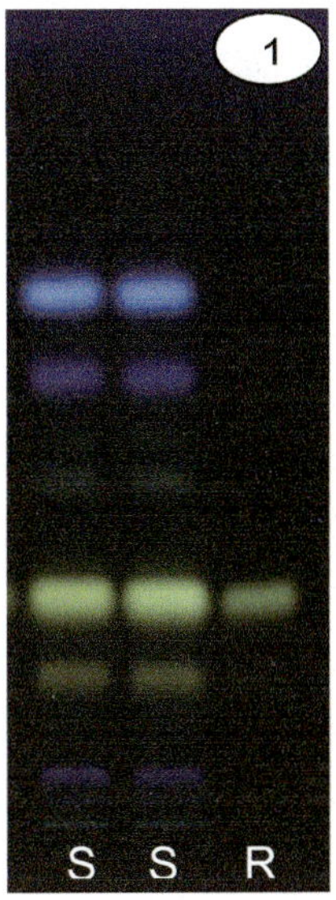

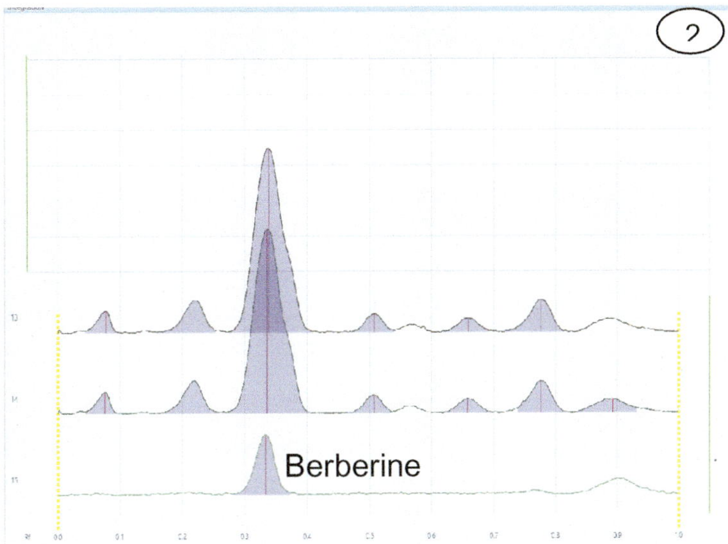

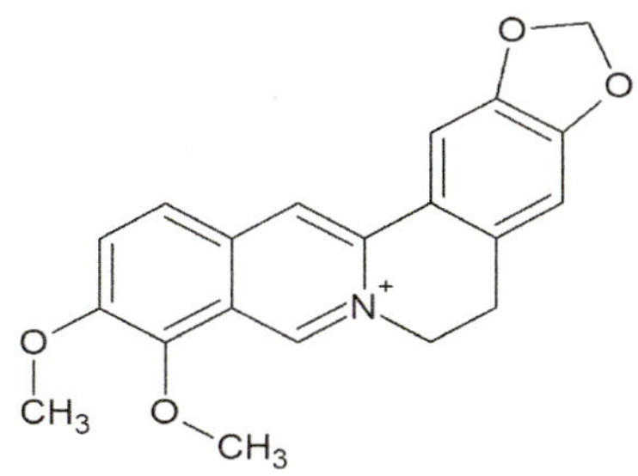

Berberine

Plate 1: HPTLC Chromatogram of *Thalictrum foliolosum* Sample (S) and Standard Berberine (R).
Plate 2: Overlay Spectra of Sample and Standard.

1. http://www.flowersofindia.net/catalog/slides/Leafy per cent 20Meadow-Rue.html.
2. Chopra RS, Chopra IC, Handa KL, Kapur L (1958). "Indigenous Drugs of India: 2nd Edition, U. S. Dhur and Sons Privated Limited, Calcutta, India: 424-425.
3. Schiff PL, Doskotch RW (1970). "*Thalictrum* Alkaloids". Lloydia, 33: 403-452.
4. Bhakuni DS, Singh RS (1982). The alkaloids of *Thalictrum foliolosum*. J Nat Pro 45(3): 252-5.
5. Pandey G, Khatoon S, Pandey MM, Rawat AK (2018). Altitudinal variation of berberine, total phenolics and flavonoid content in *Thalictrum foliolosum* and their correlation with antimicrobial and antioxidant activities. J Ayu int med 9(3): 169-76.
6. Walter NS, Bagai U (2015). Antimalarial Efficacy of *Thalictrum foliolosum* (Meadow rue) Against Chloroquine-Resistant P. falciparumr. J Trop Dis 2(163): 2.

49

Tinospora cordifolia (Thunb.) Miers

Family

Menispermaceae

Vernacular name [1-2]

Sanskrit: Amrita, Guduchi; Hindi: Giloe, Guduchi; English: Indian Tinospora; Gujarati: Galo; Bengali: Gulancha; Telugu: Tippatiga, Iruluchi; Marathi: Shindilakodi.

Distribution [3-4]

Tinospora cordifolia is found throughout the India up to an altitude of 1000 feet. The species is indigenous to India, Myanmar, Sri Lanka, China, Thailand and Bangladesh. In India it is available in Uttar Pradesh, Madhya Pradesh, Chhattisgarh, Kerala, Haryana, Tamil Nadu, Karnataka, Sikkim and Orissa.

Traditional Uses [5-6]

Tinospora cordifolia is well known Ayurvedic medicinal plant used in traditional system of medicine. Stem juice is orally taken for curing diabetes, fever, jaundice, diarrhoea, dysentery, diuretic, skin disease and to kill intestinal worm. Stem and roots are used as an antidote in snake bite and scorpion sting.

Major Chemical Constituents [7-8]

Berberine, palmitine, tinosporin and tinocordifolioside.

A Vegetative Plant

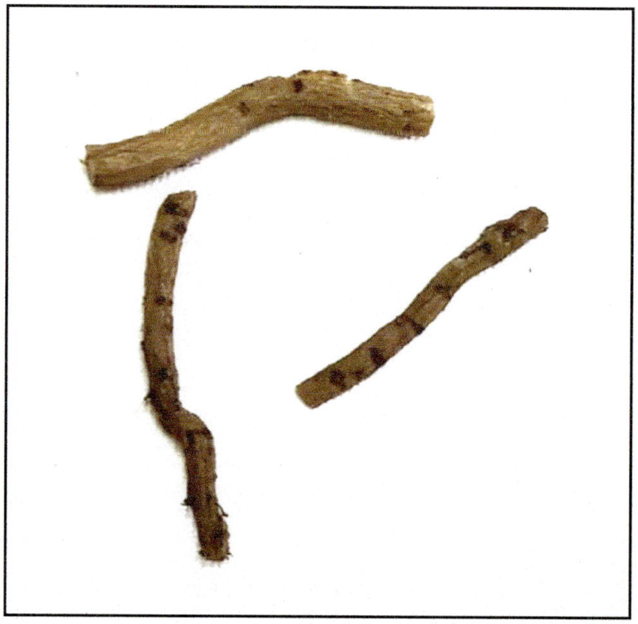

Fresh Root

Major Pharmacological Activity [9-13]

Anti-oxidant, hypoglycemic, anti-bacterial, anti-inflammatory and hepato-protective activity.

Plant Description

Tinospora cordifolia is a large perennial deciduous climbing shrub with succulent stem. Leaves are simple, heart shaped, alternate, entire, estipulate, lamina broadly ovate and multicoated reticulate venation. Stem is long, succulent, filiform, fleshy and climber. Roots are of two types, aerial and underground root. Arial roots are arises from the stem branches and characterised by tetra to penta-arch primary structure. Flowers are unisexual, small, greenish yellow, male flowers are present in cluster form whereas female flowers are solitary. Fruits are small, ovoid, aggregate, smooth, fleshy and orange-red coloured.

Anatomy

Transverse section of root is circular in outline, shows narrow cork and wider cortex, and wide vascular system, occupying major area of section. Detailed structure accounted that the root has single layered epidermis, followed by multi-layered thin-walled cork. Outer cortex is parenchymatous and inner sclerenchymatous, contains round or oval starch grains. Phloem is present just below the cortex containing sieve tube, companion cell, fibers and parenchyma. 4 – 7 vascular bundles are present in a ring and xylem is exarch with large parenchymatous pith in center. In longitudinal section xylem vessels are seen vertically running broad segmented canals with xylem and phloem fibres are vertically arranged.

HPTLC Studies

Sample Solution

Accurately weight (5 g) crude powdered sample was subjected to cold extraction with methanol (50 ml), continuously shaken for 6 h and then allowed to stand at room temperature (25 ^{0}C ± 2) for 18 h. The extraction was repeated thrice, filtered and the pooled filtrate was concentrate in rotatory evaporator (Buchi, Switzerland). Working solution of strength 10 mg/ml was freshly prepared on the day of quantative analysis.

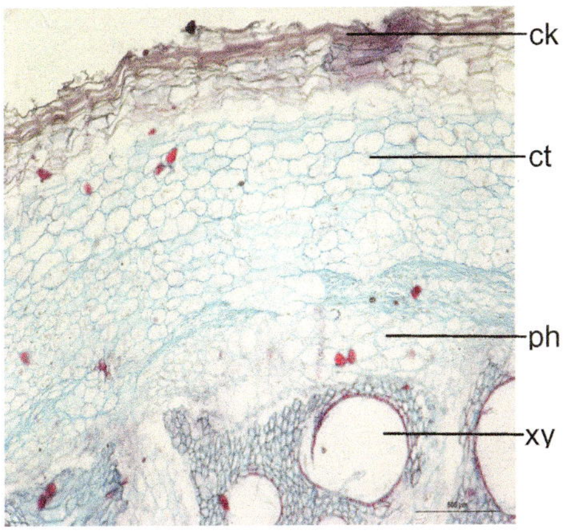

T.S. (20 x)

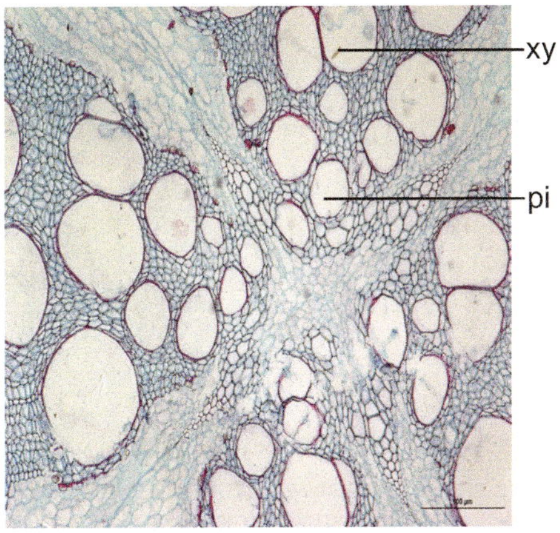

T.S. (20 x)

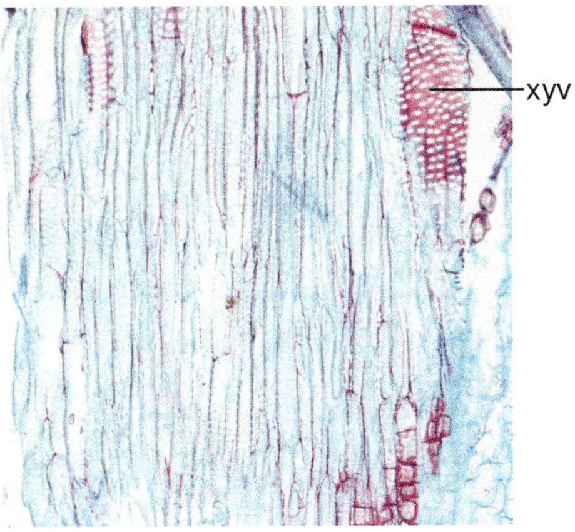

L.S. (20 x)

Transverse Section (T.S.) & Longitudinal Section (L.S) of *Tinospora cordifolia* Root.
Abbreviation:; ck- cork; ct- cortex; ph- phloem; pi- pith; xy- xylem; xyv- xylem vessels.

Standard Solution

Berberine (≥ 98 per cent) was freshly prepared of strength 1 mg/ml.

Sample Application

2 μl of standard solution and 10 μl of sample was applied as 6 mm band with CAMAG HPTLC applicator on activated silica gel 60 GF_{254} HPTLC plate. The chromatogram was developed on pre-saturated CAMAG HPTLC twin trough chamber for 10 min and the solvent system was allowed to run for 8 cm.

Solvent System

n-propanol- water- formic acid (9:1:0.1 v/v).

Visualization

The chromatogram was visualized at UV_{366} and densitometric scanning of berberine was done at 350 nm.

Identification of Berberine

The standard berberine was identified at Rf 0.32.

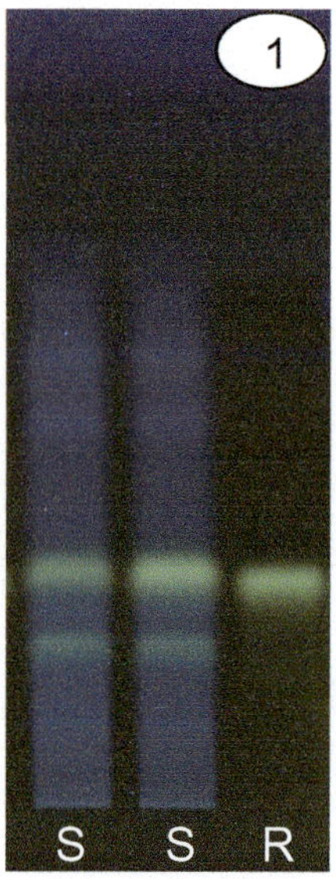

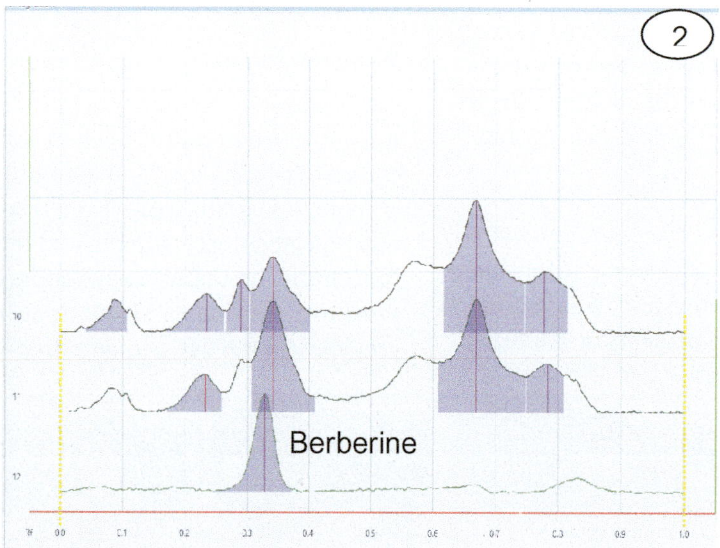

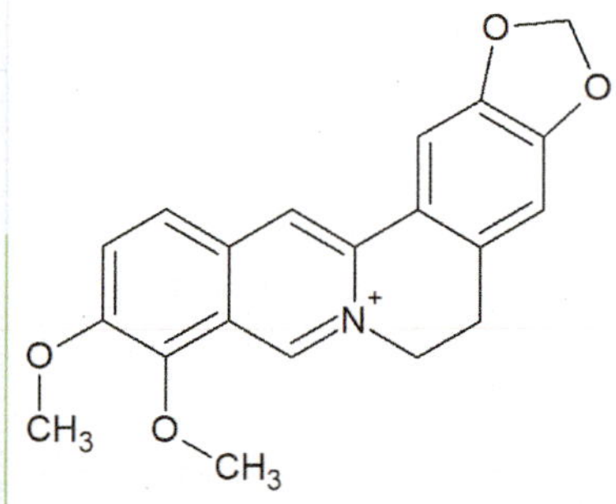

Berberine

Plate 1: HPTLC Chromatogram of *Tinospora cordifolia* sample (S) and Standard Berberine (R).
Plate 2: Overlay Spectra of Sample and Standard.

1. Sharma A, Gupta A, Singh S, Batra A (2010). *Tinospora cordifolia* (Willd.) Hook. F. and Thomson- A plant with immense economic potential. J Chem Pharm Res 2(5): 327-333.
2. Anon (1956).The Wealth of India: Raw Materials, Council of Scientific and Industrial Research, New Delhi. vol X 251–252.
3. Singh J, Sinha K, Sharma A, Mishra NP, Khanuja SP (2003). Traditional uses of *Tinospora cordifolia* (Guduchi). J Med Aromat plant Sci 25: 748-51.
4. Pramanik A, Gangopadhyay M (1993). Menispermaceae. In: Sharma BD, Balakrishnan NP, Rao RR. And Hajra PK. (eds) Flora of India 1. Calcutta: Botanical Survey of India.
5. Khory RN, Katrak NN Materia, Medica of India and their therapeutics: Komal Prakash 31.
6. Kiritikar KR, Basu BD (1933). Indian medicinal plants. Vol I Lalit Mohan Publishers, India 77.
7. Sharma A, Gupta A, Batra SSA (2010). *Tinospora cordifolia* (Willd.) Hook. F. and Thomson-A plant with immense economic potential. J chem pharm Res 2: 327-333.
8. Karpova EA, Voznyi YV, Dudukina TV, Tsvetkva IV (1991). 4-Trifluoromethylumbelliferyl glycosides as new substrates form revealing diseases connected with hereditary deficiency of lysosome glycosidases. Biochem Int 24: 1135-1144.
9. Jain S, Sherlekar B, Barik R (2010). Evaluation of antioxidant potential of *Tinospora cordifolia* and *Tinospora sinensis* Int J Pharm Sci Res 1(11): 122-8.
10. Stanely P, Prince M, Menon VP (2000). Hypoglycemic and other related actions of *Tinospora cordifolia* roots in alloxan-induced diabetic rats. J Ethnophar 70: 9-15.

11. Shanthi V, Nelson R (2013). Anitbacterial activity of *Tinospora cordifolia* (Willd) Hook. F. Thoms on urinary tract pathogens. Int J Curr Microbiol App Sci 2(6): 190-4.

12. Wesley JJ, Christina AJ, Chidambaranathan N (2008). Effect of alcoholic extract of *Tinospora Cordifolia* on acute and subacute Inflammation. Pharmacologyonline 3: 683-687.

13. Mehrotra R, Katiyar CK, Gupta AP (2000). Hepatoprotective composition and composition for treatment of conditions related to hepatitis Band E infection. US Patent, 6: 136,316.

50

Utleria salicifolia Bedd. Ex. Hook. F.

Family

Periplocaceae

Vernacular Name

English: Swallow root.

Distribution [1]

Utleria salicifolia is endemic to South Western Ghats of India (Kerala and Tamil Nadu).

Traditional Uses [1]

Malasar and Kadar tribes (Kerala), use tubers for preparing prickles for treating intestinal ailment like colic and bleeding in ulcer. The dried powder is orally taken with hot water for treatment of asthma.

Major Chemical Constituents

Lupeol, β-sitosterol and quercetin

Plant in Natural Habitat

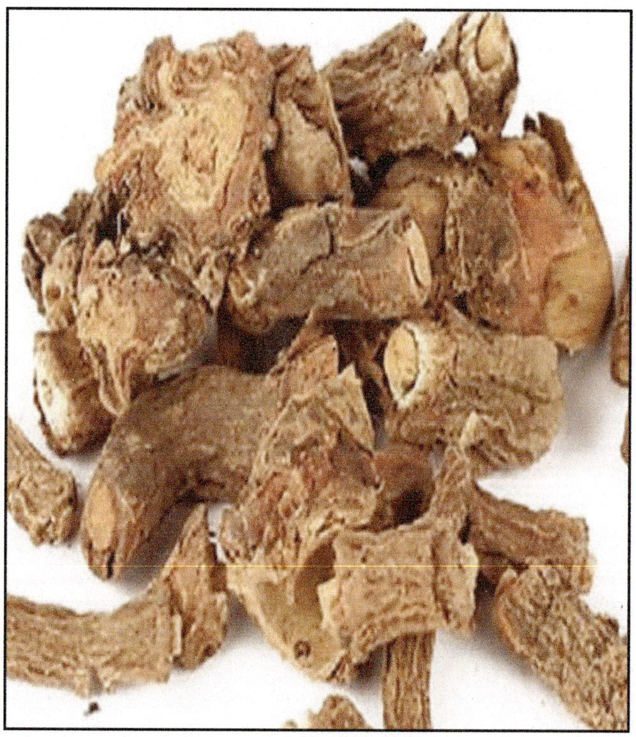

Dried Roots

Pharmacological Activity [2-4]

Antiulcer, antiproliferative and anti-inflammatory activity

Morphological Description

Utleria salicifolia is an annual or perennial short lived, erect or spreading herb growing up to 1 meter. Stem is woody, erect, and cylindrical. Leaves are small, alternate, unifoliate, petiolate, ovate-oblong or rounded, glabrous covered with dense soft hairs. Inflorescence is terminal, flowers are small, purple or white colored and seed are kidney shaped. The root is poorly developed; smooth, light yellow in colour and bark is yellowish white and has a leathery texture and bitter taste.

Anatomy

Transverse section of root is circular in outline, show very narrow range of cork, wider cortex and well developed wide stellar system. Detailed structure accounted that the root shows narrow cork, followed by multi-layered thin-walled cork cambium lie underneath this. The cortical region is made up of 15 to 18 layered irregular parenchymatous cells, containing round or oval starch grains. Just below the cortex phloem are present having sieve tubes, companion cells and phloem fibers. The xylem is 8-10 celled wide and consists of vessels, fibers, tracheids and xylem parenchyma. Vessels are mostly in groups of 5 to 8 and rarely solitary. The medullary rays are thick-walled, multiseriate, radiating, and parenchymatous containing starch grains.

HPTLC Studies

Working Solution

Accurately weigh (5 g) crude powdered was extracted (cold) with methanol (50 ml) followed by continuously shaken for 6 h and then allowed to stand for 18 h at room temperature (25 ± 2°C). The extraction was repeated thrice and the pooled filtrate was concentrate in rotatory evaporator (Buchi, Switzerland), dried residue was stored at 4 °C. Working solution of 10 mg/ml was freshly prepared on the day of quantative analysis.

Standard Solution

Lupeol (≥ 98 per cent) was freshly prepared of strength 1 mg/ml.

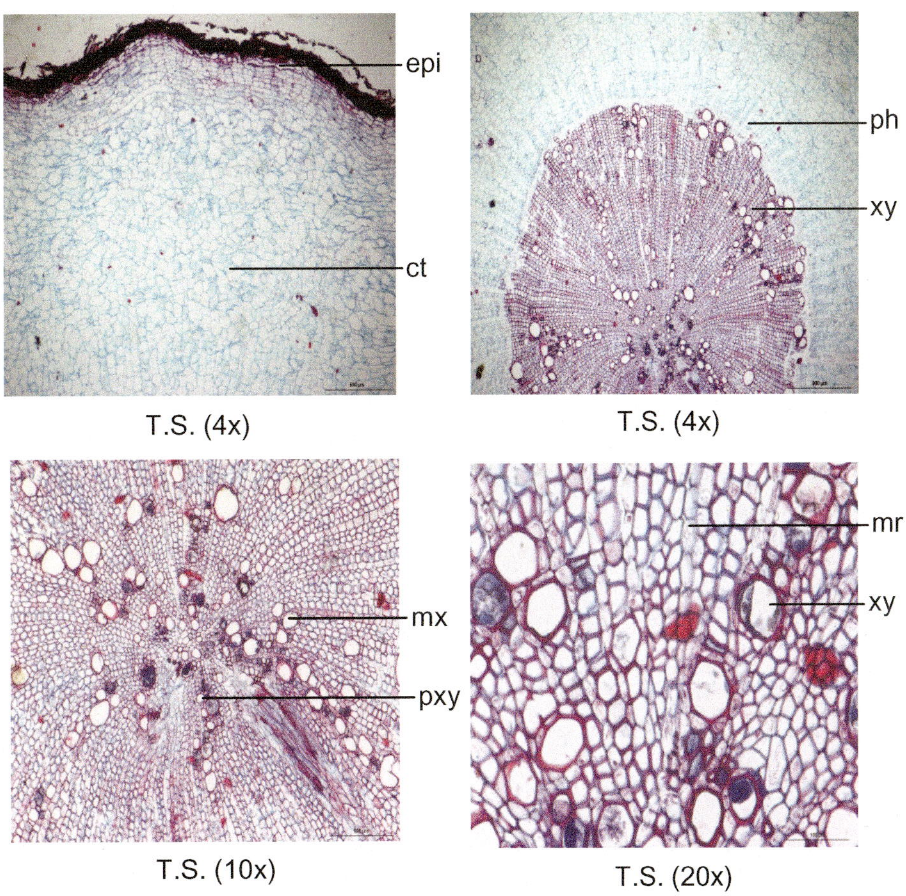

Transverse Section (T.S.) of *Utleria salicifolia* Root.

Abbreviation:; ct- cortex; epi- epidermis; mr- medullary rays; mx- meta xylem; ph- phloem; pxy- proto xylem; xy- xylem.

Sample Application

2 µl of standard solution and 10 µl of sample was applied as 6 mm band with CAMAG HPTLC applicator on activated silica gel 60 GF_{254} HPTLC plate. The chromatogram was developed on pre-saturated CAMAG HPTLC twin trough chamber for 20 min and the solvent system was allowed to run for 8 cm.

Solvent System

Toluene: ethyl acetate: methanol: formic acid (7:2:0.5:0.5 v/v).

Visualization

The chromatogram was visualized after derivatization with anisaldehyde sulphuric acid reagent and densitometric scanning was done at 530 nm.

Identification of Lupeol

The standard lupeol was identified at R_f 0.62.

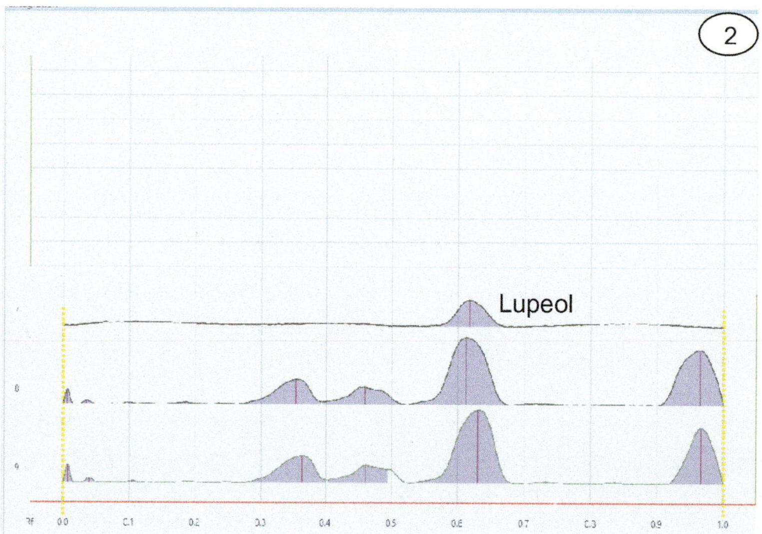

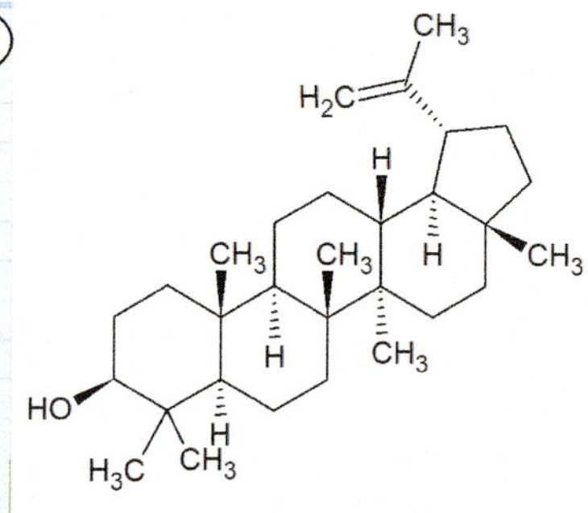

Lupeol

Plate 1: HPTLC Chromatogram of *Utleria salicifolia* Sample (S) and Standard Lupeol (R) in White Light.
Plate 2: Overlay Spectra of Sample and Standard.

1. Radhakrishnan K, Pandurangan AG, Pushpangadan P (1998). *Utleria salicifolia*—a new ethnobotanical record from Kerala, India. Fitoterapia 69: 403–405.
2. Rao CV, Ojha SK, Radhakrishnan K, Govindarajan R, Rastogi S, Mehrotra S, Pushpangadan P (2004). Antiulcer activity of *Utleria salicifolia* rhizome extract. J Ethnophar 91(2-3):243-9.
3. George S, Sebastian DP, Kavitha CV, Raghavan SC, Augustine A, Kusumanchi P (2016). In vitro antiproliferative effect of the root extracts from *Decalepis hamiltonii*, *Hemidesmus indicus* and *Utleria salicifolia*.
4. Vineetha S (2017). Antiulcer and Anti-Inflammatory Activities of *Utleria Salicifolia* Leaves in Rat Models (Doctoral dissertation, JKK Nattraja College of Pharmacy, Komarapalayam).

Botanical Terms

Actinomorphic: Relating to a flower that can be divided by any longitudinal plane into two symmetrical halves

Acuminate: Leaf with pointed, narrow, tailed apex

Adventitious root: Roots that arise from any other organ, except root

Alternate: The leaves are single at each node and arise in an ascending spiral along the stem

Anther: Part of the stamen contains pollen

Bark: Outermost layer of stem and root of woody plants

Berry: A small, pulpy, and often edible fruit

Bract: A modified or specialized leaf associated with a reproductive structure such as a flower, inflorescence axis or cone scale

Cambium: Partially undifferentiated cells help in plant growth

Collateral: Vascular bundle containing xylem and phloem adjacent to each other

Companion cells: Companion cells are closely coupled with sieve element; it regulates the activity of sieve element and participates in loading and unloading of sugar

Cork: Non-living cells; made up of waxy substance suberin

Cortex: Tissues of parenchymatous cells located between the epidermis and vascular tissues of stem and root

Creeper: Plants growing through spreading stems or branches along the ground or other surface

Dithecous: Tetrasporangiate anther having four macrosporangium

Endodermis: Endodermis is the central, innermost layer of cortex. The endodermis helps to control the circulation of water, ions and hormones in the vascular system

Entire leaf: Entire leaves having a smooth margin without incisions

Epidermis: Epidermis is single layer of cells that cover leaves, flowers, roots and stems. It forms a boundary between the plant and the external environment

Erect: Rigidly upright or straight

Evergreen: Plants whose leaves are green and functional in more than one season

Exstipulate: Leaves without stipules

Flower: Reproductive part of a flowering plant

Fibrous root: It is usually made up of thin, medium branching roots growing from the stem, mostly found on Monocots

Gamopetalous: Petals united

Glabrous: Leaf without any kind of hair

Globose: Spherical

Ground tissue: Ground tissue of plants includes all tissues that are neither dermal nor vascular

Gynandrous: Stamens and pistil united

Herbaceous: Vascular plants that have soft woody stem

Hermaphrodite: Any organism having both male and female sex organs

Heterogeneous: More than one type of cells

Horizontal: Parallel to the plane of the horizon

Hypogynous: Having the stamens and other floral parts situated below the carpels

Imbricate: Out of five petals, one is completely internal and one is external

Imparipinnate: Single leaflet at the apex of a compound leaf

Intercellular space: Vacant space between cells

Internode: Length of stem between two nodes

Lanceolate: Leaves shaped like a lance head

Leaflet: Small leaf like part of a compound leaf

Leaf scars: The mark left on stem after a leaf falls

Margin dentate: Toothed margin of leaf

Medullary rays: Medullary rays are cellular structures found in wood. They appear as radial planar structures, perpendicular to the growth rings

Metaxylem: Xylem formed after the protoxylem, vessels large in diameter

Node: Point of attachment of leaf to primary stem

Obovate: Egg-shaped leaf with narrow end at base

Oleoresin: A natural or synthetic mixture of essential oils and resin

Opposite: Two leaves arise from the stem at the same node, on opposite sides

Ovate: Egg-shaped

Ovary: Female organ of a flower; it is enlarged basal portion of the pistil

Ovoid: More or less egg-shaped.

Parenchymatous cells: Cellular tissue, usually soft and succulent; loosely packed cells, intercellular spaces are present among the cells

Peduncle: Stalk supporting an inflorescence

Perennial: Perennial plant completing their life cycle in more than two years

Perianth: Non-reproductive part of flower, consisting calyx and corolla

Pericycle: Outer most part of the vascular bundle of plants

Petal: Modified leaves that surround the reproductive parts of the flower

Petiole: A stalk that attaches leaf to stem

Petiolate: Petiole present

Phloem: Phloem is a living part of the vascular system, transferring soluble organic materials and foods

Phloem fibers: It supports the phloem's conductive cells and provides strength to the stem

Polyarch: Having many protoxylem groups

Polygonal: Cells with flat faces and straight edges; having many sides

Proto xylem: First xylem formed from procambium

Pubescent: Presence of leaf hair

Raceme: Unbranched, indeterminate type of inflorescence bearing pedicellate flowers along its axis

Reticulate venation: Veins shows net like arrangement all over the leaf blade

Rhizome: Rhizome is a type of plant that is either located on the soil surface or underground which consists of nodes from which roots and seedlings grow

Rootlets: Rootlets are defined as small branches of roots formed from the main roots

Rosette crystals: Rosette crystals are compound crystals formed from the aggregation of small crystals arranged radially around a central core

Sclereids: Sclereids diminish as sclerenchymatous cells, highly dense, lignified cell walls

Secondary growth: Secondary growth is the result of cell division in the lateral meristems that thickens the stem and root

Sepal: The outermost whorl of flower parts, protects the petals as they develop

Sessile: Petiole is absent

Shrub: Small to medium sized perennial woody plant having persistent woody stem

Sieve tubes: It is a specialized cell of phloem that helps in the transport of organic compounds formed during photosynthesis

Solitary: Alone, single

Spike inflorescence: It is similar to a raceme inflorescence, but bearing sessile flowers

Stamen: Pollen-producing reproductive organ of a flower

Stele: In vascular plants, stele is the central part of root or stem containing the tissues derived from procambium

Stigma: The head of the pistil

Sub sessile: Petiole is very short

Tendril: The tendril is a specialized stem, leaf or petiole with a threadlike shape, used by climbers for support and cellular invasion by parasitic plants

Tracheids: Tracheids are elongated cells in the xylem that helps in the transport of water and mineral salts

Trichome: Tiny hairs either unicellular or multicelluler, that emerge from the surfaces of leaves and stem of plants

Trifoliate: A leaf divided into three leaflets

Tuber: It is modified stem used in storage of food in certain seed plants

Unisexual: Flowers representing only one sex either male or female

Vascular bundle: Vascular bundle is a part of the transport system in vascular plants, which is present in two components: xylem and phloem

Xylem: Xylem is a non-living part of the vascular system; it carries water and dissolved minerals from the roots to the rest of the plant and also provides mechanical support

Xylem parenchyma: Xylem parenchyma is mainly involved in the storage of carbohydrate, fats and water conduction

Xylem fiber: Any fiber composed of dead sclerenchyma cells between the xylem vessels and tracheids of xylem tissue

Xylem vessels: Xylem vessels are long tubes that help transport water and provide mechanical support. It is formed from end to end union of large number of small, wide, lignified dead cells

Unifoliate: Leaf that never divides into small leaflets

Wood: Wood is a porous and fibrous structural tissue found in the stem and root of woody plant

Zygomorphic: Flower that can be divided by single longitudinal plane into two symmetrical halves

Therapeutic Terms

Abortifacient: A substance that induces abortion

Alzheimer's disease: A progressive disease that destroys memory and other important mental functions

Analgesic: Painkiller

Anemia: A condition in which the blood does not have enough healthy red blood cells

Anorexia: An eating disorder

Anthelmintic: It is used to expel parasitic worms and other internal parasites from the body

Anti-amoebic: Drugs used in treatment of infection caused by amoebae

Anti-bacterial: A medicine used to prevent the growth or spread of bacteria

Anti-convulsant: Drugs that inhibit neural activity

Antidote: A substance that can counteract form of poisoning

Antidiabitic: A drugs used to stabilize and control blood glucose level amongst people with diabetes

Anti-diarrheal: Drugs used in the treatment of diarrhea

Anti-fertility: Drugs that have the ability to reduce fertility

Anti-fungal: Drug or Medications used to treat fungal infections

Anti-gout: These agents work to correct overproduction or underexcretion of uric acid

Anti-hyperlipidemic: Drugs or chemical substance used in reduction of lipid levels in blood

Antihypertensive: Drugs that are used to treat hypertension

Anti-inflammatory: Treatment that reduces inflammation or swelling

Anti-migratory: Chemical substance used in inhibiting migration

Anti-obesity: Any medication that reduces or control weight

Antioxidants: Antioxidants are compounds that prevent oxidation, a chemical reaction that produces free radicals that may damage the cells of organism

Anti-plasmodial: A drug which is effective against apicomplexan parasites in the genus Plasmodium

Anti-proliferative: Chemical substance used to inhibit cell growth

Anti-pyretic: A drug that reduces fever

Anti-septic: Antiseptics are antimicrobial substances that are applied to living tissue/skin to reduce the chance of infection

Anti-thrombotic: A drug that reduces the formation of blood clots

Anti-tumour: Drugs that inhibit the growth of a tumour

Anti-viral: Drug or medication used for treating viral infections

Anxiety: Mental health disorder characterized by feelings of worry or fear

Aphrodisiac: It is a substance that increases sexual desire, sexual pleasure, or sexual behavior

Asthma: A condition in which a person's airways become inflamed, narrow and swell and produce extra mucus, makes it difficult to breathe

Astringent: Astringent is a chemical that shrinks or constricts body tissues

Bronchitis: Inflammation in the tubes that carry air to your lungs

Cardiotonic: Drugs that have beneficial action on the heart

Carminative: A drug that relieves flatulence

Conjunctivitis: Inflammation or infection of the outer membrane of the eyeball and inner eyelids

Depurative: Substances used in purifying and detoxifying the blood

Diaphoretic: The power to increase sweating

Diarrhea: Loose, watery feces that occur more often than usual, caused by virus and sometimes contaminated food

Diuretic: Any substance that increase the production of urine

Dropsy: Inflammation of soft tissues due to accumulation of excess water

Dysentery: Inflammation of the intestine with bloody diarrhea

Dysmenorrhea: The term dysmenorrhea is used to describe painful periods

Dysuria: Discomfort or pain when urinating

Eczema: An itchy inflammation of the skin

Elephantiasis: Parasitic infection that causes excessive swelling in the arms and legs

Emollient: It makes skin soothe and hydrate

Epilepsy: A disorder in which nerve cell activity in the brain is disturbed

Excitement: A feeling of great excitement and eagerness

Expectorant: A medication that helps to bring up mucus and other contents from the lungs, bronchi and trachea

Febrifuge: Medicine used to reduce fever

Fever: Body temperature higher than usual

Gonorrhea: A sexually transmitted disease that may cause infertility

Gout: A form of arthritis that causes severe joint pain

Haemorrhoids: Swollen and inflamed veins in the rectum and anus that cause discomfort and bleeding

Headache: Pain in the head or upper neck of the body

Hemostasis: A process to prevent and stop bleeding from damaged blood vessels

Hepato-protective: The ability of a chemical substance to prevent damage to the liver

Hypertension: A medical condition in which the blood pressure in the arteries steadily increases

Hypothermia: A condition in which body loses heat faster than it produces

Immunomodulatory: Drugs that modifies functioning of the immune system

Immunosuppressive: Partially or completely suppressing the immune response

Inflammation: It is the biological response of body tissues to harmful stimuli, such as pathogens or damaged cells

Insecticidal: Insecticides are substances used to kill insects

Insomnia: A sleep disorder in which people have trouble sleeping

Jaundice: Yellow skin due to formation of bilirubin in blood

Larvicidal: An insecticide that is specifically targeted against the larval life stage of an insect

Leprosy: A chronic, curable infectious disease mainly causing skin lesions and nerve damage caused by Mycobacterium leprae

Leucoderma: A disease that causes loss of skin colour and form spots

Leucorrhea: A whitish or yellowish vaginal discharge

Leukemia: It is a blood cancer caused by an increase in the number of white blood cells in your body

Malaria: A deadly disease caused by a plasmodium parasite, spread by the bite of infected mosquitoes

Neuroprotective: The ability of a therapy to prevent neuronal cell death

Parkinson's disease: It is a central nervous system disorder that affects movement, balance and coordination

Plague: A rare but serious bacterial infection that is spread by fleas

Pimple: Red, soft bumps with white pus on the face

Pneumonia: Pneumonia is an infection that inflames the air sacs of lungs, causes cough and fever

Psoriasis: A condition in which skin cells build up and form scales and dry patches

Pungent: Having a strong taste or smell

Purgative: They are used for the treatment and prevention of constipation

Rheumatism: It is a condition that causes chronic, often intermittent pain affecting the joints or connective tissue

Roundworms: Roundworms are parasites that live in your intestine

Scabies: A contagious, intensely itchy skin condition caused by a tiny, burrowing mite

Spermicidal: A contraceptive pills that destroys sperm

Sprains: A stretch or tear of the ligaments

Stomachache: Pain in a person's stomach or belly.

Swelling: Enlargement of organs, skin or other parts of the body

Syphilis: A bacterial infection is usually spread through sexual contact

Tonic: A traditional drink used in the treatment of various diseases

Toothache: Pain or inflammation in the tooth gum

Tuberculosis: A bacterial disease that primarily affects the lungs

Tympanitis: Inflammation in middle ear

Typhoid: Typhoid fever is spread by contaminated food and water, characterized by high fever, diarrhea and vomiting

Ulcer: A sore that develops on the lining of the oesophagus, stomach, or small intestine

Urethritis: inflammation of the urethra; tube that carries urine from the bladder to outside the body

Vasorelaxant: Chemical substance that relax the vessels constriction

Wounds: Injuries that break the skin or other body tissue

Index

B

C

D

E